Innovative Biomolecular Structure Analysis Techniques

Innovative Biomolecular Structure Analysis Techniques

Guest Editor

Tzanko I. Doukov

MDPI

Basel • Beijing • Wuhan • Barcelona • Belgrade • Novi Sad • Cluj • Manchester

Guest Editor
Tzanko I. Doukov
SLAC National Accelerator Laboratory
Stanford University
Stanford, CA
USA

Editorial Office
MDPI AG
Grosspeteranlage 5
4052 Basel, Switzerland

This is a reprint of the Special Issue, published open access by the journal *Biomolecules* (ISSN 2218-273X), freely accessible at: https://www.mdpi.com/journal/biomolecules/special_issues/ND19IA513D.

For citation purposes, cite each article independently as indicated on the article page online and as indicated below:

Lastname, A.A.; Lastname, B.B. Article Title. *Journal Name* **Year**, *Volume Number*, Page Range.

ISBN 978-3-7258-7865-9 (Hbk)
ISBN 978-3-7258-7866-6 (PDF)
https://doi.org/10.3390/books978-3-7258-7866-6

Contents

About the Editor

Tzanko I. Doukov

Tzanko I. Doukov is a scientist and part of the Macromolecular Crystallographic Group at the Stanford Synchrotron Radiation Lightsource, Menlo Park, CA 94025, USA. His research focuses on macromolecular crystallography, metalloproteins, structural dynamics, and the study of protein structures at physiological temperatures.

biomolecules

MDPI

Article

Crystallography of Extremophile Proteins—Structural Comparisons of Psychrophilic and Hyperthermophilic Rubredoxins

Tzanko Doukov [1,*], Trenton F. Turpin [2], Dominic George [3], Caroline Cole [2], Kat Drumright [4], Madigan Rumley [5], Ryan Boyce [6], Francis E. Jenney, Jr. [6] and Stephen P. Cramer [7,*]

1 SLAC National Laboratory, SSRL, Menlo Park, CA 94025, USA
2 University of Texas, Austin, TX 78712, USA; trentonturpin@utexas.edu (T.F.T.); mary@carolinecole.org (C.C.)
3 University of Saskatchewan, Saskatoon, SK S7N 5A2, Canada; dag488@mail.usask.ca
4 University of California, Davis, CA 95616, USA; ksdrumright@ucdavis.edu
5 University of Colorado-Boulder, Boulder, CO 80309, USA; madigan.rumley@gmail.com
6 BIomedical Science, Philadelphia College of Osteopathic Medicine, Georgia Campus, Suwanee, GA 30024, USA; ryanbo2@pcom.edu (R.B.); francisje@pcom.edu (F.E.J.J.)
7 SETI Institute, Mountain View, CA 94043, USA
* Correspondence: tdoukov@slac.stanford.edu (T.D.); spjcramer@seti.org (S.P.C.)

Abstract

Psychrophilic organisms are able to grow at temperatures down to −15 °C, while hyperthermophiles can multiply at temperatures up to 122 °C. What structural changes in extremophile proteins are needed to maintain stable and biochemically active structures under such conditions? Understanding how such extremophiles accomplish this is relevant for human health, biotechnology, and our search for life elsewhere in the universe. The purpose of the current study is to report and compare the structures of four rubredoxins (Rds), the first ever two experimental psychrophile bacteria structures (from Gram-positive *Clostridium psychrophilum* and Gram-negative *Polaromonas glacialis*) and two hyperthermophiles from the Gram-negative *Thermotoga maritima* bacterium and the archaeon *Pyrococcus yayanosii*, also a piezophile, as part of a program to understand structural variations that support both stability and function under extreme conditions. These structures were obtained using synchrotron radiation X-ray diffraction at 100 K. All four structures had the expected overall rubredoxin fold. Rubredoxin from the only aerobic psychrophilic bacterium *Polaromonas glacialis* had larger variations in sequence and structure, whereas the other psychrophilic bacterium showed properties closely related to hyperthermophile rubredoxins. Multi-subunit structures showed similar RMSD variability independent from their thermal adaptation status. We propose including functional information in the analysis since temperature optimization may not be the only determinant for a specific protein adaptation.

Keywords: rubredoxin; extremophile; hyperthermophile; psychrophile; X-ray diffraction

1. Introduction

For life to thrive under extreme conditions [1,2], its constituent proteins need to maintain stable and biochemically active structures [3–6]. Understanding the structure of extremophile proteins under such conditions is interesting in its own right, and it is also relevant for human health, biotechnology [7], and our search for life elsewhere in the universe [8]. The same is even true for proteins from 'non-extremophiles': how does structure change as a function of temperature and other conditions?

One approach to explaining extremophile protein stability is the 'corresponding states' proposal, which posits that proteins have evolved to achieve comparable flexibility at the optimal growth temperature for their respective organisms [9,10]. In support of this idea, neutron scattering experiments found comparable flexibility for psychrophiles and thermophiles at their respective adaptation temperatures (from 4 to 85 °C) [11]. However, some experiments conflict with its predictions. For example, using NMR-monitored amide hydrogen exchange experiments, hyperthermophilic *Pyrococcus furiosus* rubredoxin (*Pf* Rd) was found to have greater flexibility compared to the mesophile *Clostridium pasteurianum* (*Cpa* Rd) protein [12,13]. In a similar vein, from neutron scattering, NMR, and other measurements, hyperthermophilic P450 CYP119 was found to be more flexible than its mesophilic counterpart CYP101A at all temperatures above 200 K [14].

In a previous study using ^{57}Fe measurements, we found comparable flexibility for hyperthermophilic *Pf* Rd compared to psychrophilic *Polaromonas glacialis* (*Pg*) Rd [15]. An additional surprise from our *Pf* Rd studies was the observation of a conformational change at around 343 K that produced modifications in hydrogen bonding [16]. However, it remains to be seen if *Pf* Rd represents a unique case, or if high temperature conformational changes are a general feature of extremophile proteins. As a starting point for addressing this issue, here we determined four new Rd structures from psychrophilic *Clostridium psychrophilum* (*Cpsy*) and *Pg*, as well as hyperthermophilic *Thermotoga maritima* (*Tm*) and hyperthermophilic and piezophilic *Pyrococcus yayanosii* (*Py*). Along with similar protein sequences (Table 1), we found comparable 100 K structures. These formed the basis for future studies at room temperature and ~373 K when possible.

Table 1. Sequences for rubredoxins with structures solved in this paper or previously aligned by *CLUSTAL Omega* [17]. [1] *Pseudomonas oleovorans* 2Fe Rd C-terminal starting at residue 119 [18,19], [2] *Mycobacterium tuberculosis* Rd 2 [20], [3] *Pseudomonas aeruginosa* Rd [21], [4] *Clostridium pasteurianum* Rd [22,23], [5] *Pyrococcus abyssi* Rd [24], and [6] *Pyrococcus furiosus* Rd. Growth temperatures are taken from the BacDive database [25]. Conserved cysteines—C, prolines—P, other prolines—**P**, and core aromatics—W, Y, and F. The rooted phylogenetic tree of the alignment is included in Supplemental Figure S1.

Organism	Sequence	Growth Temp °C	PDB ID
Cpsy	MNKYVCLVCGYEYDPEIGDLEGGIKPGTKFEDLPEDWLCPLCGVTKFDFEKI	4	9ZDO
Pg	-MTWMCLICGWIYDEALGSPEHGIAAGTPWSQVPMNWTCPECGARKEDFEMVQM	10	9ZDP
Po ct [1]	YLKWICITCGHIYDEALGDEAEGFTPGTRFEDIPDDWCCPDCGATKEDYVLYEEK	30	1A24
Mt 2 [2]	YKLFRCIQCGFEYDEALGWPEDGIAAGTRWDDIPDDWSCPDCGAAKSDFEMVEVARS	37	7A9A
Pa [3]	MKKWQCVVCGLIYDEAKGWPEEGIEAGTRWEDVPEDWLCPDCGVGKLDFEMIEIG	37	2V3B
Cpa [4]	MKKYTCTVCGYIYNPEDGDPDNGVNPGTDFKDIPDDWVCPLCGVGKDQFEEVEE	37	1FHH
Tm	MKKYRCKLCGYIYDPEQGDPDSGIEPGTPFEDLPDDWVCPLCGASKEDFEPVE	80	9ZDI
Pab [5]	MAKWRCKICGYIYDEDEGDPDNGISPGTKFEDLPDDWVCPLCGAPKSEFERIE	95	1YK5
Py	MAKWRCTVCGYIYDEEEGDPDNGVLPGTKFEELPDDWVCPLCGAPKDMFEKVD	98	9ZDH
Pf [6]	MAKWVCKICGYIYDEDAGDPDNGISPGTKFEELPDDWVCPICGAPKSEFEKLED	100	5NW3
#	000000000111111111122222222223333333333444444444455555		
#	123456789012345678901234567890123456789012345678901234		

Yellow boxes highlight non-conserved Prolines, while cyan boxes highlight partially conserved Asp-19 and Leu-41.

2. Materials and Methods

2.1. Protein Preparation

The extremophile Rds were expressed in *Escherichia coli*. Recombinant vectors containing the genes that code for these rubredoxins were synthesized with codon optimization [26] for expression in *E. coli* in plasmid pET24d by GenScript (Piscataway, NJ, USA). We used *Cpsy* Rd (Genbank accession number WP_216289245.1), *Tm* Rd (Genbank accession number AKE30317.1), and *Py* Rd (Genbank accession number AEH24664.1). Production of *Pg* Rd (Genbank accession number WP_198026828.1) has been previously described [15].

The plasmids were electroporated into *E. coli* T7Express (New England Biolabs, Ipswich, MA, USA) and maintained in LB medium with 50 ug/mL kanamycin. The recombinant Rds were expressed and purified essentially as previously described [16] with a few modifications. *Py* Rd was purified the same way except that all three N-terminal forms (N-fMet, N-Met, and N-Ala (the native form)) were separated on hydroxyapatite chromatography. The *Tm* and *Cpsy* Rds were judged pure after gel filtration chromatography. Zinc forms of the recombinant proteins were prepared as described [16]. All protein solutions were concentrated to 40 mg/mL, 50 mM Tris pH 8.0, and 300 mM NaCl.

2.2. Protein Crystallization

All crystals were obtained with the hanging drop technique in Linbro plates with various NaCl concentrated solutions in the well. The specific conditions were: (a) Zn *Cpsy* Rd: Each drop contained 2 μL of protein solution, mixed with 2 μL 2.0 M $NaKHPO_4$ and 1 μL *Cpsy* protein crystal derived seeds. The well contained 0.5 mL of 75% NaCl. (b) Fe *Pg* Rd: Each drop contained 2 μL of protein solution diluted to 14 mg/mL in 0.22 mM NAD, 50 mM Tris pH 8.0, 300 mM NaCl, and 2 μL of 2 M $(NH_4)_2SO_4$ solution over 0.5 mL of a 70% saturated NaCl well solution. (c) Fe *Py* Rd: The coverslip contained a drop produced from 2 μL distilled water added to 2 μL protein solution, and 4 μL of 3.2 M $NaKHPO_4$ and 100 mM Tris pH 7.0. Drops were equilibrated vs. a completely empty well. (d) Zn *Tm* Rd: Each drop contained 2 μL of protein solution, mixed with 2 μL 4.0 M $(NH_4)_2SO_4$ and 1 μL *Tm* protein crystal derived seeds. The coverslip was equilibrated versus 0.5 mL of 80% saturated NaCl well solution.

2.3. Crystal Mounting

Protein crystals were harvested at the beamlines under ParatoneN oil, and the surface water layer was removed using a Hampton nylon loop (Hampton Research, Aliso Viejo, CA, USA). Finally, before mounting on the goniometer, the excess oil was dabbed away.

2.4. Diffraction Data Collection

All diffraction data were collected at SSRL. Data for a needle Zn *Cpsy* Rd crystal (1.2 × 0.15 × 0.10 mm) were collected using the helical data collection mode across the long side of the needle at BL12-2 at 0.7293 Å, allowing for a final data resolution of 0.84 Å in space group $P2_12_12_1$ (sg19) with a single Rd molecule per asymmetric unit. Despite the wavelength being far from the Zn edge (1.2836 Å), it was possible to phase *de novo* the structure with HKL2MAP [27]. A dataset at 1.7711 Å on the same crystal was used to confirm the presence of 3 K ions based on their anomalous signal.

An Fe *Pg* Rd crystal (150 × 100 × 50 μ) was used for data collection at 0.9795 Å at BL12-2 with 5% transmission, a 40 × 40 μ beam, and a calculated absorbed dose of 4.5 MGy. The strong anomalous signal from the Fe atoms allowed for *de novo* phasing with HKL2MAP [27] in space group $P4_32_12$ (sg96), resulting in 7 unique Rd molecules in the asymmetric unit.

https://doi.org/10.3390/biom16050623

A single Fe *Py* Rd crystal (10 × 100 × 50 μ) dataset was collected at 0.9795 Å at BL12-2. Molecular replacement with an Fe *Pf* Rd model (1BRF) was used to locate the 3 subunits in the asymmetric unit of space group $P2_12_12_1$ (sg19). A weak anomalous signal from the Fe ions confirmed the metal center positions but was insufficient for *de novo* phasing.

A Zn *Tm* Rd crystal (150 × 150 × 100 μ) dataset was collected at 0.7749 Å at BL12-2 at a 1.02 Å data resolution in space group $P2_1$ (sg4). In addition, a second dataset from the same crystal at 1.2826 Å, corresponding to the Zinc peak wavelength from the MAD scan, was used for *de novo* phasing, resulting in 2 Rd molecules in the asymmetric unit.

2.5. Diffraction Data Processing

Diffraction data recorded on an Eiger 16M PAD detector (DECTRIS AG, Baden-Daettwil, Switzerland) [28] was processed with the XDS (19 January 2025) [29], Pointless (1.13.4) [29,30], Aimless (0.8.2) [31], CCP4 (9.0.008) [32], and STARANISO (2.4.16) programs, as implemented in the autoPROC software (1.0.5) [33].

2.6. Protein Sequence and Structure Analysis for Temperature Adaptation

Protein temperature adaptation, as well as all other properties, is encoded in the amino acid sequence. A pairwise sequence identities table was produced by BLASTp (2.17.0) [34], indicating the various degrees of difference for the novel and already known Rds in Table S1. Multisequence alignment was performed and a rooted phylogenetic tree was created using CLUSTAL Omega (1.2.4) [17] for Table 1 and Figure S1. Protein amino acid content was analyzed using ProtParam (EXpasy) [35,36]

Structural analysis was based on the output from the BANDIT web server for normalizing the B-factors per dataset [37]. Rigidity analysis was performed using the ProPHet rigidity web server [38]. Protein void volume and density calculations were produced using the ProteinVolume web server [39]. Surface charges were calculated in Pymol (3.1.4.1) [40]. New structures were analyzed using the ProteinTools server [41] and Amino Acid Interactions (INTAA) web server v2.0 [42].

3. Results

3.1. Overall Structures

The overall folds for these four new Rd structures were quite similar to the previous structure for *Pf* Rd [16], as illustrated in Figure 1. The structures all contain the essential features of the 'consensus' Rd, including the knuckles, β-sheets, aromatic cores, and loops A and B.

The Zn *Cpsy* Rd single monomer 0.84 Å structure was determined in space group $P2_12_12_1$ with R/R_{free} = 14.0%/15.8% and contained 71 waters and 3 K atoms. The Fe *Pg* Rd structure was determined in space group $P4_32_12$ with seven independent Rd monomers to 1.83 Å, R/R_{free} = 17.2%/20.1%, and 227 identified waters and one Na ion. The Fe *Py* Rd structure with three independent Rd monomers was determined in space group $P2_12_12_1$ to 1.36 Å with R/R_{free} = 17.4%/21.2% and with 209 waters and a Na ion. The Zn *Tm* Rd structure with two independent Rd monomers was determined in space group $P2_1$ with R/R_{free} = 15.3%/18.1% and with 114 identified waters. The 100 K crystal structures for these four Rds have been submitted to the Protein Data Bank with the following PDB IDs: (a) Zn *Cpsy* Rd, 9ZDO; (b) Fe *Pg* Rd, 9ZDP; (c) Fe *Py* Rd, 9ZDH; and (d) Zn *Tm* Rd, 9ZDI. Experimental statistics for the crystallography experiments are listed in Table 2.

https://doi.org/10.3390/biom16050623

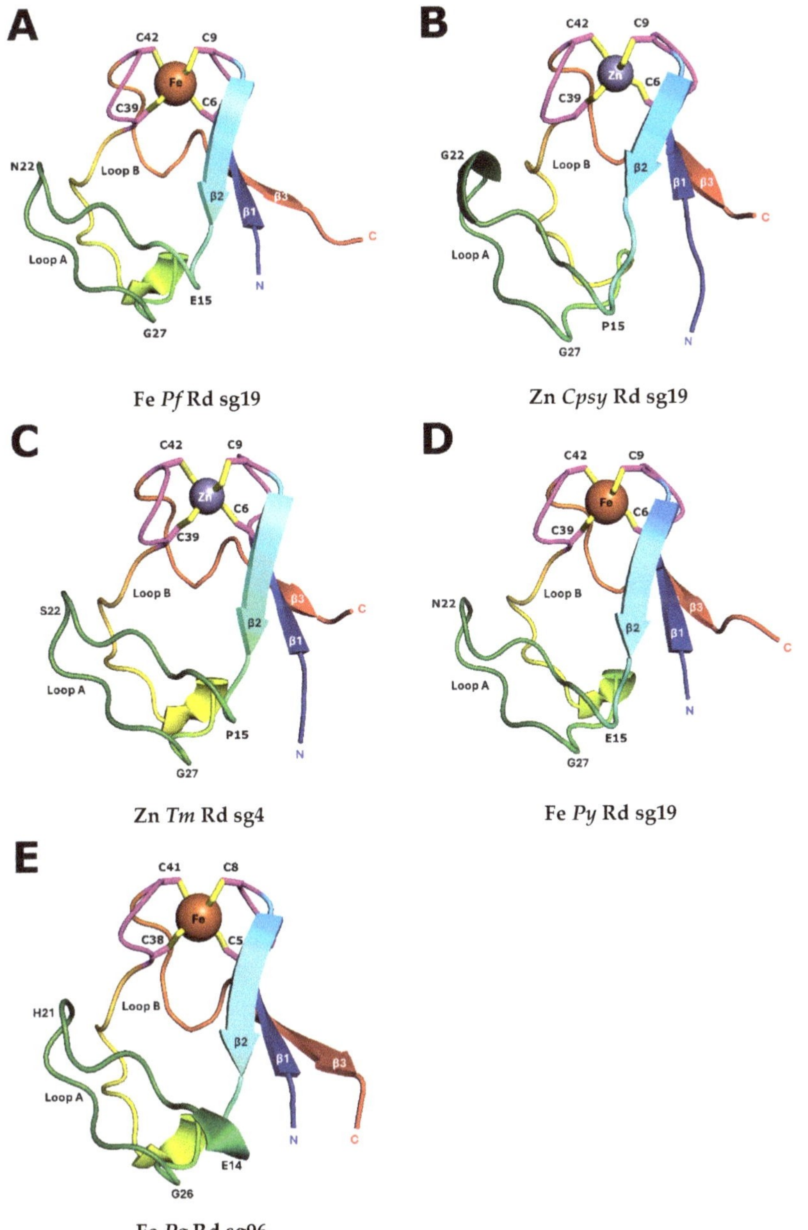

Figure 1. Rainbow cartoon representation for the structures for: (**A**) Fe *Pf* Rd, (**B**) Zn *Cpsy* Rd, (**C**) Zn *Tm* Rd, (**D**) Fe *Py* Rd, and (**E**) Fe *Pg* Rd. The figure was created with *Pymol* [40]. The label sg refers to the space group number.

Table 2. Data collection, processing, and refinement statistics for the four new Rd crystal structures at 100 K.

Protein	Zn *Cpsy* Rd	Fe *Pg* Rd	Fe *Py* Rd	Zn *Tm* Rd
PDB ID	9ZDO	9ZDP	9ZDH	9ZDI
Temperature (K)	100	100	100	100
SSRL beamline	BL12-2	BL12-2	BL12-2	BL12-2
Crystal size (mm)	1.20, 0.15, 0.10	0.15, 0.10, 0.05	0.01, 0.10, 0.05	0.15, 0.15, 0.10
Wavelength (Å)	0.72929	0.97946	0.97946	0.77488
Resolution range (Å)	27.39–0.84 (0.95–0.84)	69.29–1.83 (2.04–1.83)	33.25–1.36 (1.50–1.36)	25.26–1.02 (1.08–1.02)
Space group	$P2_12_12_1$	$P4_32_12$	$P2_12_12_1$	$P2_1$
Unit cell: a, b, c (Å) α β γ	27.48, 37.02, 40.72 90 90 90	73.90, 73.90, 199.35 90 90 90	26.91, 63.51, 78.06 90 90 90	27.02, 50.03, 33.93, 90 110.8 90
Total reflections	231,110 (9614)	1,951,770 (86,408)	175,948 (8462)	209,291 (10427)
Unique reflections	25,522 (1276)	36,053 (1803)	17,219 (861)	34,060 (1703)
Multiplicity	9.1 (7.5)	54.1 (47.9)	10.2 (9.8)	6.1 (6.1)
Completeness (spherical) (%)	66.5 (10.9)	73.0 (13.5)	57.8 (11.4)	79.1 (25.6)
Completeness (ellipsoidal) (%)	91.0 (50.6)	95.3 (69.9)	89.7 (57.5)	88.0 (50.5)
Mean I/sigma(I)	9.3 (1.9)	14.1 (1.6)	6.4 (1.5)	6.2 (1.8)
Wilson B-factor (Å^2)	11.941	20.709	20.306	14.477
R-merge	0.115 (1.56)	0.277 (6.053)	0.433 (3.708)	0.205 (3.317)
R-pim	0.040 (0.838)	0.038 (0.862)	0.138 (1.216)	0.089 (1.427)
$CC_{1/2}$	0.997 (0.515)	0.998 (0.608)	0.979 (0.837)	0.985 (0.340)
ISa	17.97	14.82	14.0	9.99
Mosaicity	0.51	0.11	0.14	0.16
Refinement				
Refinement range	27.39–0.84 (0.95–0.84)	69.29–1.83 (1.88–1.83)	33.25–1.36 (1.39–1.36)	25.26–1.02 (1.04–1.02)
R_{work} (%)	14.0 (31.6)	17.2 (29.8)	17.4 (27.6)	15.3 (26.6)
R_{free} (%)	15.8 (0.00)	20.1 (51.3)	21.2 (29.7)	18.1 (31.8)
No. of non-H atoms				
Total	537	3153	1480	970
Macromolecules	466	2918	1257	854
Ligands (Zn/Fe, Na/K)	1 Zn, 3 K	7 Fe, 1 Na	3 Fe, 1 Na	2 Zn
Water	71	227	209	114
R.m.s. deviations				
Bond lengths (Å)	0.012	0.007	0.008	0.011
Angles (°)	2.240	1.704	1.589	1.953
Average B-factor (Å^2)	17.134	47.743	17.478	13.718
Macromolecules	15.841	47.778	15.915	12.652
Ligands (Zn/Fe, Na/K)	4.21 Zn, 18.78 K (3)	39.8 Fe(7), 34.0 Na	9.33 Fe(3), 26.3 Na (1)	8.1 Zn (2)
Water	26.104	47.293	26.931	21.276

https://doi.org/10.3390/biom16050623

Table 2. *Cont.*

Protein	Zn *Cpsy* Rd	Fe *Pg* Rd	Fe *Py* Rd	Zn *Tm* Rd
Clashscore	3.19	1.79	1.64	0.6
MolProbity score	1.72	0.94	0.91	0.7
Ramachandran Plot				
Favored (%)	94	100	100	100
Allowed (%)	6	0	0	0
Outliers (%)	0	0	0	0

3.2. Metal Coordination

The metal coordination distances and angles confirmed the expected pseudo tetrahedral coordination by four cysteine residues with slightly longer bond lengths for the Zn-substituted proteins. Table S2 lists all the metal–ligand distances in the Rd molecules. Ligands shielded by aromatic residues Cys6 (by Phe49) and Cys39 (by Tyr11) have longer bonds to the metal center, but in some of the structures this general rule was not obeyed. It would be instructive to investigate with other spectroscopy methods if this is a real effect of part of ensemble sampling on crystal contacts within the multi-subunit crystals, or if it is due to lower crystal resolution.

3.3. Rd Fold Determinants on Molecular Level

The Rd fold is commonly described as a combination of the two knuckles around the iron and three antiparallel β strands. The structure is also held by additional specific bonds, which are also the most mechanically rigid parts of the molecules and the most resistant to H/D exchange. These include the already mentioned C6, C9, C39, and C42 coordinating the metal, the Y13 side chain with a T28 backbone, and the tight β turn backbone bond between K46 and F49 (*Pf* Rd numbering).

3.4. Asp-19 H-Bonding

In our previous study of hyperthermophile *Pf* Rd [16], at 100 K we observed H-bonding from Asp19 to Trp37, with a connection to Tyr11 through two water molecules (Figure 2A). At 260 K there was only a single water bridge to Tyr11 (Figure 2B), and at 393 K Asp19 rotated to make a direct H-bond to Tyr11 (Figure 2C) [16]. For comparison, the mesophile *Cpa* Rd structure 1FHH at 100 K is analogous to the observed 260 K Zn *Pf* Rd structure with a single bridging water to Tyr11 (Figure 2D) [23].

The H-bonding pattern for Asp19 varies considerably in our new set of low-temperature Rd structures. In the psychrophilic *Cpsy* structure, the Asp19 H-bond network looks quite like 100 K *Pf* Rd, with a direct H-bond from Asp19 to Trp37 and a linkage to Tyr11 through two water molecules. It only lacks the stabilizing additional hydrogen bond from Asn22 in *Pf* Rd, since *Cpsy* Rd has Gly22 instead (Figure 2F). Both hyperthermophile structures have an Asp19 network conserved on a sequence level. The *Tm* structure also shows direct Asp19 to Trp37 H-bonding, but the connection to Tyr11 is more tenuous, with some water molecules at ~3.3 Å (Figure 2G). An interesting variability pattern is observed in *Py* Rd, where in subunit A the Asp19 keeps a similar hydrogen bond pattern to Trp37 (Figure 2H), while in subunit B it utilizes a bridging water molecule to simultaneously reach the Trp37 ring as well as the Tyr11 ring (Figure 2I). Surprisingly, the Asp19 is ~4.6 Å and ~3.9 Å away from both Trp37 and Tyr11 in subunit C without a well-ordered density in any bridging positions (Figure 2J). Note that corresponding amino acids in the *Pf* and *Py* sequences occupy three different spatial positions depending on the subunit in the unit cell (Figure 2A,H–J). Finally, in *Pg* Rd, there is an Ser18 (−1 shift for all residues compared to

https://doi.org/10.3390/biom16050623

the rest of the sequences) in the place of Asp19, and a Trp10 in the place of Tyr11, which lacks the hydrogen bond network (Table 1) (Figure 2E).

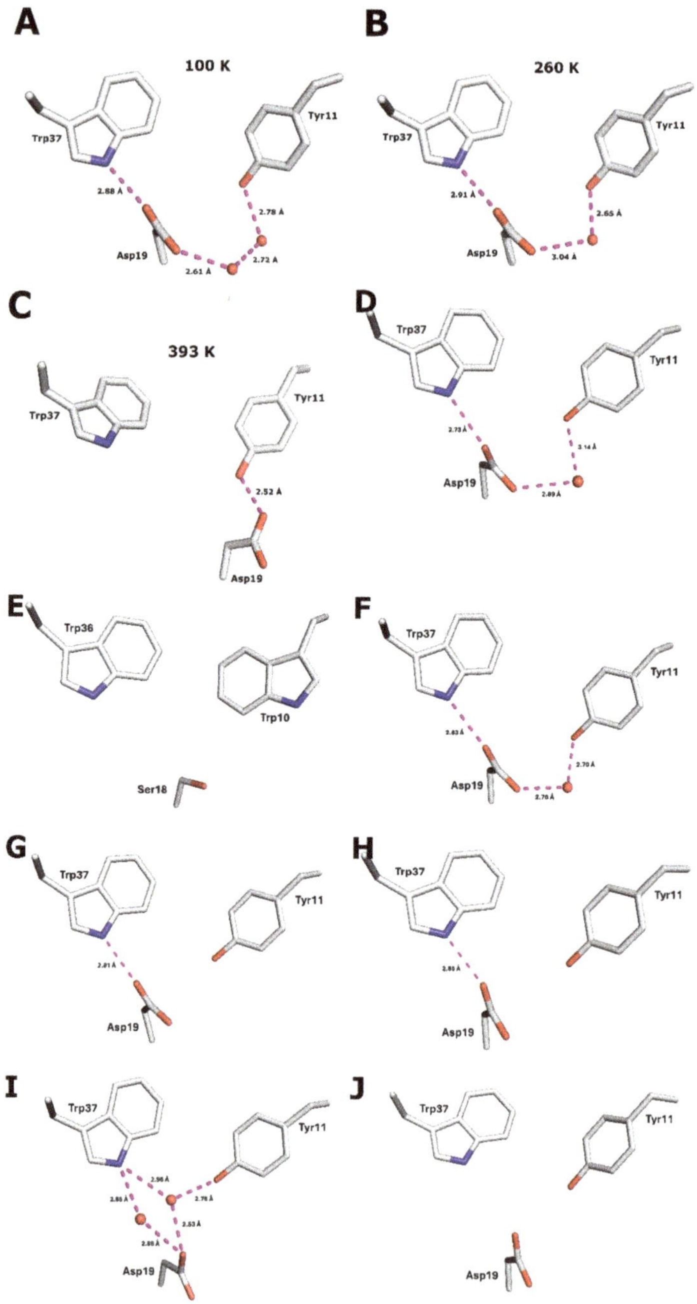

Figure 2. H-bonding patterns around Asp19: (**A**) *Pf* Rd at 100 K. (**B**) *Pf* Rd at 260 K. (**C**) *Pf* Rd at 393 K. (**D**) *Cpa* Rd at 100 K. (**E**) *Pg* Rd at 100 K. (**F**) *Cpsy* Rd at 100 K. (**G**) *Tm* Rd at 100 K. (**H**) *Py* Rd—subunit A at 100 K. (**I**) *Py* Rd—subunit B at 100 K. (**J**) *Py* Rd—subunit C at 100 K.

https://doi.org/10.3390/biom16050623

3.5. Glu15 H-Bonding

Extensive H-bonding around Glu15 has often been invoked as a source of stability for the hyperthermophile *Pf* Rd [43–45]. In all of the *Pf* Rd structures, Glu15 interacts with the indole ring of Trp4 through a long hydrogen bond and the amide O of Phe30 and the N-terminal Ala2 (Figure 3A). In *Py* Rd (Figure 3B) the Glu15 hydrogen bonding pattern is conserved. In contrast, in psychrophile *Cpsy* Rd (C) and hyperthermophile *Tm* Rd (D), as well as in the mesophile *Cpa* Rd, there is a Pro at position 15, disrupting the hydrogen bonding network. Finally, in *Pg* Rd, the analogous residues are Glu14, and Trp3 and Met1 (Arg5 H-bonds in ZnTmRd—Figure 3E), but the H-bonds to the N-terminal residue are too long or in an unfavorable geometry in four (subunits B, E, F, and G) of the seven subunits, making the H-bonding network requirement very flexible/complex depending on the structural content inside the crystal.

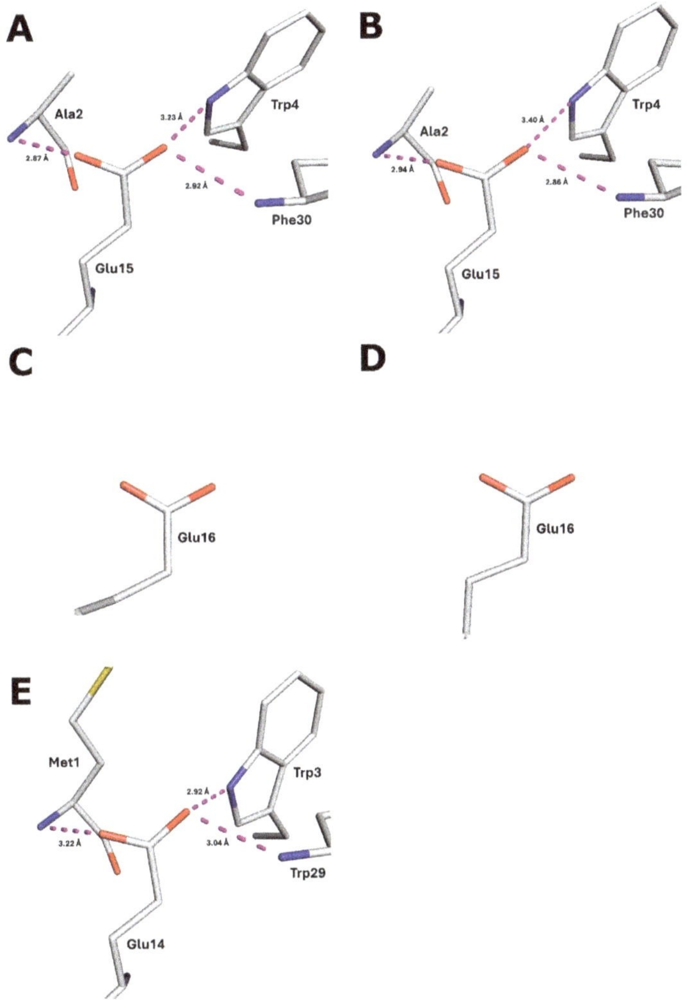

Figure 3. H-bonding patterns around Glu15 at 100 K. (**A**) *Pf* Rd (100 K). (**B**) *Py* Rd. (**C**) *Cpsy* Rd. (**D**) *Tm* Rd. (**E**) *Pg* Rd.

https://doi.org/10.3390/biom16050623

3.6. Arg5 -Glu50 Salt Bridge in ZnTmRd

Tm Rd lost the favorable and energy stabilizing Glu15 hydrogen bond network but remained a hyperthermophile. How could that be possible? Careful examination of the amino acids' interaction energies using the INVAA server showed that a new favorable salt bridge was formed between Arg5 and Glu50. It was present in both subunits of the Zn *Tm* Rd structure as illustrated in Figure 4.

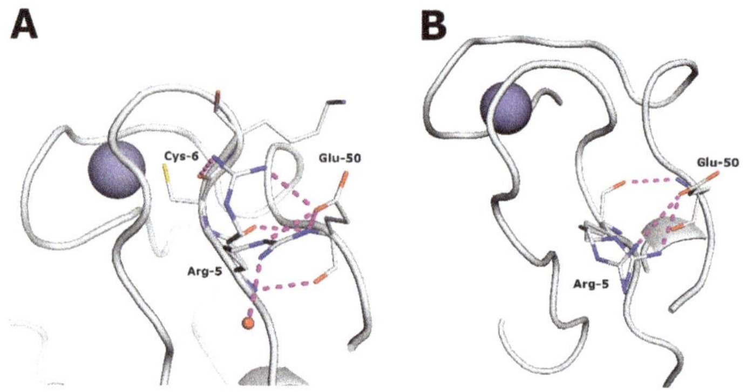

Figure 4. H-bonding patterns around Arg5 at 100 K in Zn *Tm* Rd structure. (**A**) Subunit A. (**B**) Subunit B.

3.7. Water Molecules

The number of water molecules observed per protein molecule is included in Table S3. We found this mostly reflects the resolution for each structure, rather than the differential hydration of the proteins.

3.8. Sequence and Structure Analysis of Temperature Adaptation

A general theory of cold adaptation is that increased molecular flexibility and decreased stability result in higher enzymatic activity [9,46]. Analyzing electron transfer psychrophilic enzymes could be very complex and challenging. This complex, multifactorial adaptation process involves fitting different functional partners to accommodate aerobic or anaerobic species and varying pressure requirements, whereas the electron transfer process itself does not rely on major structural changes. [15].

Sequences of the novel bacterial and archaeal rubredoxins were compared with previously determined rubredoxin structures (Table 1). A phylogenetic tree of that comparison (Figure S1) indicates *Polaromonas glacialis* rubredoxin belongs to the aerobic branch, suggesting a different functional role in aerobic metabolism. Based on multiple protein families with psychrophilic, mesophilic and thermophilic X-ray structure representatives, Feller summarized the psychrophilic protein adaptations as: (a) an increase in glycine residues favoring higher flexibility, (b) a decrease in proline in loop regions, which usually are rigidifying protein structures, (c) a decrease in arginine residues which also rigidify structures by salt bridges and/or hydrogen bonds, and (d) a reduction in the size of the amino acids in the hydrophobic core, therefore weakening the stabilizing hydrophobic forces [47]. Additionally, the role of increased negative surface charge in enabling flexibility at low temperatures was suggested [47]. Data based on these criteria is listed in Table 3. Psychrophilic *Cpsy* and *Pg* rubredoxins have unexpectedly small negative surface charges.

Table 3. Amino acid content (K, R, D, E, G, and P) and properties for the novel and previously characterized rubredoxins. Unusually small surface charges for the psychrophiles are marked as **bold**.

Protein	#aa	#atoms	Hydrophobic	Acidic	Basic	Neutral	#K/R	#D+E	Surface(−)	#G	#P
Cpsy	52	817	38.46	21.15	9.62	30.77	5/0	11	**−4.99**	6	4
Pg	53	806	47.17	13.21	5.66	33.96	1/1	7	**−5.04**	5	4
Po	55	848	32.73	25.45	9.09	32.73	3/1	14	−8.99	5	3
Mt2b	60	905	40.00	23.33	8.33	28.33	2/3	14	−8.99	5	3
Pa	55	857	41.82	23.64	9.09	25.45	4/1	13	−6.99	7	3
Cpa	54	808	31.48	24.07	7.41	37.04	4/1	13	−7.99	5	7
Tm	53	808	35.85	24.53	9.43	30.19	4/1	13	−7.99	5	7
Pab	53	816	35.85	24.53	11.32	28.30	4/2	13	−6.99	5	5
Py	53	812	39.62	24.53	9.43	26.42	4/1	13	−7.99	5	5
Pf	54	817	38.89	24.07	9.26	27.78	5/0	13	−7.99	5	5

"Determination of molecular flexibility is complex as it requires the definition of the types and amplitudes of atomic motions as well as a timescale for these motions." [47]. Preliminary circular dichroism spectra for the psychrophile *Pg* Rd indicate melting temperatures (>50 °C) at the β-sheet wavelengths in the thermophilic temperature range (manuscript in preparation). On a structural level, the very wide RMSD subunit distribution (Table S4), the high average B-factor (~47, Table 2), and the B-factor distribution (Figures S3–S5) for the *Pg* subunits indicate intrinsic flexibility. Secondary structure assignment by the STRIDE algorithm [48] also showed variations in the overall very conserved fold (Table S5). Even in thermophilic rubredoxins (Figures S2–S4) with multiple subunits, variations were observed in flexible loop B, suggesting that some of the structural information may be content-dependent on crystal contacts, solvent content differences, and/or precipitant concentration and nature (inorganic salt or PEG).

Local structural rigidity could be identified computationally with the ProPHet rigidity server. While the thermophilic proteins tend to have higher absolute peaks (>120 for Y13), one of the psychrophiles, *Cpsy* Rd, with a low solvent content also has a similar peak height, suggestive of content-dependent modulation of the properties (Figure S5).

Another way to destabilize the proteins is having larger cavities and voids in the psychrophile proteins. *Cpsy* Rd and some of the *Pg* Rd subunits have lower protein densities, but some of the *Pg* subunits have densities higher than mesophile or thermophile proteins in the table (Table S6). Therefore, using that criterion to identify temperature adaptation is not advisable. *Pab* Rd data illustrates that protein density is lower at room temperature than at 100 K. The *Pf* Rd crystal (PDB 5NW3) exhibits the highest protein density, owing to its lower solvent content and higher resolution.

The amino acid interactions server (INTAA) derives interaction energies between amino acids. Its concept was verified by an ab initio study of rubredoxin [49]. The current version of the server incorporates both detailed interaction energy calculations and amino acid conservation within an interaction energy matrix (IEM). A convenient way to evaluate the overall state of the protein is a Scatter Plot of the interaction energy versus information content (IC) as a measure of amino acid conservation within multisequence alignments. It brought to our attention the strong stabilizing effect of Arg5 in *Tm* Rd, an amino acid not conserved universally in rubredoxins. Total IE-IC plots are included in Figure S6.

https://doi.org/10.3390/biom16050623

4. Discussion

In this work we presented a set of four new Rd crystal structures—two from hyperthermophiles and, for the first time, two from psychrophiles. All four structures follow the conserved Rd fold and metal center coordination, and they all include a six-aromatic-residue core. However, there are some interesting differences.

On a sequence level, the Rd from the psychrophilic *Pg* bacterium (the only aerobe in the set) differs significantly from the other three with the lack of D19 (here S18) H-bond networks, as well as substitutions within the core aromatics (W10 instead of Y11 and W29 instead of F30) (Table S1). Its second amino acid in the Fe binding motif CXXC is E40, compared to the most common L41 (*Cpa*, *Cpsy*, *Py*, and *Tm*) or I41 (*Pf*). We speculate that the charged carboxylate residue from the glutamate could lead to a different redox potential for *Pg*, compared to the anaerobic organisms. Supporting this idea, other Rds from aerobes have a D at position 41, and the charged carboxylate from the aspartate should have a similar effect. When compared to known Rd sequences (Table 1), *Pg* Rd fits into the RubB (Rd type 2) family which includes *Mt* Rd [20] and *Pa* Rd [21], both of which are involved in dioxygen chemistry with Cytochrome P450 or Alkane monooxygenase (AlkB). Although the redox partners of *Pg* Rd have not been assigned, both P450 and AlkB are present and seem likely candidates.

The other psychrophilic protein, *Cpsy* Rd, contains a rare seventh aromatic amino acid. Phe47 is located on a solvent-exposed loop, where it hinders stabilizing hydration of the protein. Although aromatic residues strongly enhance stability, Phe47's contribution is the smallest of all *Cpsy* aromatic residues. Its B-factors are roughly three times higher than the B-factors of the buried aromatic residues. The closest aromatic residue is symmetry-related Tyr11 at ~4.9 Å. *Cpsy* Rd has Pro15 instead of the stabilizing hydrogen-bonded Glu15 found in *Pf* rubredoxin. This is reflected in the observation of multiple conformations for the aromatic residue at position 30 (Phe30), which is usually very ordered in other rubredoxins (e.g., Trp29 in *Pg* Rd).

Organisms utilize multiple molecular mechanisms to adapt to cold environments [50], increasing metabolic activity despite reaction rates that are unfavorable according to the Arrhenius Law. It is proposed that the general mechanism involves increasing flexibility by disrupting stabilizing interactions. Such molecular changes lead to decreased protein stability [51]. There is not a universal molecular mechanism for cold adaptation. Adaptations vary between different protein groups. Because rubredoxin is a very small electron transfer protein, its molecular adaptation signals are subtle, making it difficult to differentiate between psychrophiles and thermophiles. From our analysis we found that only decreased negative surface charge correlates with expected lower thermal stability in rubredoxins (Table 3). Rubredoxins exhibit a melting temperature significantly higher than the optimal growth temperature of the organism, presenting an unexpectedly large discrepancy. Surprisingly, both psychrophilic and thermophilic proteins showed significant structural variations across multiple crystal subunits. Instead of acting as solid, uniform units, thermophilic proteins displayed varied B-factor profiles.

5. Conclusions

We reported X-ray diffraction rubredoxin structures from the anaerobic hyperthermophile bacterium *Thermotoga maritima* and piezophile archaeon *Pyrococcus yayanosii*. We also reported for the first time the experimental structures of two psychrophilic rubredoxins—from anaerobe *Clostridium psychrophilum* and aerobe *Polaromonas glacialis* bacteria. Their overall structures were remarkably preserved within the rubredoxin fold with three antiparallel β-strands and four cysteine–iron binding sites. We applied multiple protein sequence analyses and structure-based tests to identify the likely origin of tempera-

https://doi.org/10.3390/biom16050623

ture adaptation. Aside from an unexpected reduction in calculated surface charge in both psychrophiles, no clear molecular clues explained the adaptation from psychrophiles to mesophiles and hyperthermophiles.

While analyzing crystals with multiple subunits, we stumbled across the significant role that crystal contacts, solvent content, data collection temperature, and resolution play upon structural malleability. This affected not only the supposedly more flexible psychrophiles but also the supposedly more rigid thermophiles. Performing future experiments at physiological temperature and pressure conditions would allow for a better capture of molecular signatures related to temperature adaptation.

Supplementary Materials: The following supporting information can be downloaded at: https://www.mdpi.com/article/10.3390/biom16050623/s1, Table S1: Pairwise sequence identity; Table S2: Metal–ligand distances; Table S3: Observed water molecules; Table S4: RMSD values for main chain atoms of different subunits; Table S5: Automated secondary structure assignment by STRIDE; Table S6: Protein Volume Server results; Figure S1: Rooted phylogenetic tree of Table 1 rubredoxin sequences; Figure S2: B-factor distribution; Figure S3: Normalized B-factor distribution for main chain atoms; Figure S4: Normalized B-factor distribution for all atoms; Figure S5: ProPHet rigidity web server results; Figure S6: Total interaction energy (IE) versus information content (IC) scatter plots.

Author Contributions: Conceptualization, S.P.C., F.E.J.J. and T.D.; methodology, T.D.; formal analysis, T.D.; investigation, F.E.J.J., T.D., D.G., M.R., R.B., C.C., K.D. and T.F.T.; resources, S.P.C.; data curation, T.D.; writing—original draft preparation, S.P.C. and T.D.; writing—review and editing, S.P.C. and T.D.; visualization, T.F.T. and T.D.; funding acquisition, S.P.C. and F.E.J.J. All authors have read and agreed to the published version of the manuscript.

Funding: This research was funded by the National Science Foundation, grant number MCB-2149122, and National Institutes of Health, grant GM-65440.

Institutional Review Board Statement: Not applicable.

Informed Consent Statement: Not applicable.

Data Availability Statement: The original data presented in the study are openly available in the Protein Data Bank (PDB) database at https://www.rcsb.org/ under the following PDB IDs: (a) Zn *Cpsy* Rd, 9ZDO; (b) Fe *Pg* Rd, 9ZDP; (c) Fe *Py* Rd, 9ZDH; and (d) Zn *Tm* Rd, 9ZDI.

Acknowledgments: We thank the National Institutes of Health grant GM-65440 (S.P.C.) and National Science Foundation grant MCB-2149122 (S.P.C. and F.E.J.J.) for supporting this work. Use of the Stanford Synchrotron Radiation Lightsource, SLAC National Accelerator Laboratory, was supported by the U.S. Department of Energy, Office of Science, Office of Basic Energy Sciences under Contract No. DE-AC02-76SF00515. The SSRL Structural Molecular Biology Program is supported by the DOE Office of Biological and Environmental Research, and by the National Institutes of Health, National Institute of General Medical Sciences (P30GM133894). The contents of this publication are solely the responsibility of the authors and do not necessarily represent the official views of NIGMS or NIH.

Conflicts of Interest: The authors declare no conflicts of interest.

Abbreviations

The following abbreviations are used in this manuscript:

Cpsy	*Clostridium psychrophilum*
Pg	*Polaromonas glacialis*
Tm	*Thermotoga maritima*
Py	*Pyrococcus yayanosii*
Pf	*Pyrococcus furiosus*
Cpa	*Clostridium pasteurianum*
Po	*Pseudomonas oleovorans*

https://doi.org/10.3390/biom16050623

Mt	*Mycobacterium tuberculosis*
Pa	*Pseudomonas aeruginosa*

References

1. Harrison, J.P.; Gheeraert, N.; Tsigelnitskiy, D.; Cockell, C.S. The Limits for Life under Multiple Extremes. *Trends Microbiol.* **2013**, *21*, 204–212. [CrossRef] [PubMed]
2. Ando, N.; Barquera, B.; Bartlett, D.H.; Boyd, E.; Burnim, A.A.; Byer, A.S.; Colman, D.; Gillilan, R.E.; Gruebele, M.; Makhatadze, G.; et al. The Molecular Basis for Life in Extreme Environments. *Annu. Rev. Biophys.* **2021**, *50*, 343–372. [CrossRef]
3. Feller, G. Protein Folding at Extreme Temperatures: Current Issues. *Semin. Cell Dev. Biol.* **2018**, *84*, 129–137. [CrossRef]
4. Panja, A.S.; Maiti, S.; Bandyopadhyay, B. Protein Stability Governed by Its Structural Plasticity is Inferred by Physicochemical Factors and Salt Bridges. *Sci. Rep.* **2020**, *10*, 1822. [CrossRef]
5. Pinney, M.M.; Mokhtari, D.A.; Akiva, E.; Yabukarski, F.; Sanchez, D.M.; Liang, R.B.; Doukov, T.; Martinez, T.J.; Babbitt, P.C.; Herschlag, D. Parallel Molecular Mechanisms for Enzyme Temperature Adaptation. *Science* **2021**, *371*, eaay2784. [CrossRef]
6. Timr, S.; Madern, D.; Sterpone, F. Protein Thermal Stability. In *Computational Approaches for Understanding Dynamical Systems: Protein Folding and Assembly*; Strodel, B., Barz, B., Eds.; Progress in Molecular Biology and Translational Science; Academic Press: Cambridge, MA, USA, 2020; pp. 239–272.
7. Harris, K.L.; Thomson, R.E.S.; Strohmaier, S.J.; Gumulya, Y.; Gillam, E.M.J. Determinants of Thermostability in the Cytochrome P450 Fold. *Biochim. Biophys. Acta-Proteins Proteom.* **2018**, *1866*, 97–115. [CrossRef]
8. D'Amico, S.; Claverie, P.; Collins, T.; Georlette, D.; Gratia, E.; Hoyoux, A.; Meuwis, M.A.; Feller, G.; Gerday, C. Molecular Basis of Cold Adaptation. *Philos. Trans. R. Soc. London. Ser. B Biol. Sci.* **2002**, *357*, 917–925. [CrossRef]
9. Závodszky, P.; Kardós, J.; Svingor, A.; Petsko, G.A. Adjustment of Conformational Flexibility Is a Key Event in the Thermal Adaptation of Proteins. *Proc. Natl. Acad. Sci. USA* **1998**, *95*, 7406–7411. [CrossRef] [PubMed]
10. Jaenicke, R. Protein Folding: Local Structures, Domains, Subunits, and Assemblies. *Biochemistry* **1991**, *30*, 3147–3161. [CrossRef]
11. Tehei, M.; Franzetti, B.; Madern, D.; Ginzburg, M.; Ginzburg, B.Z.; Giudici-Orticoni, M.T.; Bruschi, M.; Zaccai, G. Adaptation to Extreme Environments: Macromolecular Dynamics in Bacteria Compared in Vivo by Neutron Scattering. *EMBO Rep.* **2004**, *5*, 66–70. [CrossRef] [PubMed]
12. Hernández, G.; Jenney, F.E.; Adams, M.W.W.; LeMaster, D.M. Millisecond Time Scale Conformational Flexibility in a Hyperthermophile Protein at Ambient Temperature. *Proc. Natl. Acad. Sci. USA* **2000**, *97*, 3166–3170. [CrossRef]
13. Hernández, G.; LeMaster, D.M. Reduced Temperature Dependence of Collective Conformational Opening in a Hyperthermophile Rubredoxin. *Biochemistry* **2001**, *40*, 14384–14391. [CrossRef] [PubMed]
14. Liu, Z.; Lemmonds, S.; Huang, J.; Tyagi, M.; Hong, L.; Jain, N. Entropic Contribution to Enhanced Thermal Stability in the Thermostable P450 Cyp119. *Proc. Natl. Acad. Sci. USA* **2018**, *115*, E10049–E10058. [CrossRef] [PubMed]
15. Jenney, F.E.; Wang, H.; George, S.; Xiong, J.; Guo, Y.; Gee, L.; Marizcurrena, J.J.; Castro-Sowinski, S.; Staskiewicz, A.; Yoda, Y.; et al. Temperature-Dependent Iron Motion in Extremophile Rubredoxins—No Need for 'Corresponding States'. *Sci. Rep.* **2024**, *14*, 12197. [CrossRef] [PubMed]
16. Doukov, T.; Leontyev, I.; Jenney, F.E., Jr.; George, D.; Cramer, S.P. Some Like It Hot—X-Ray Diffraction at 120 °C Reveals Structural Changes in Extremophile Protein. *Angew. Chem. Int. Ed.* **2025**, *65*, e20302. [CrossRef]
17. Madeira, F.; Madhusoodanan, N.; Lee, J.; Eusebi, A.; Niewielska, A.; Tivey, A.R.N.; Lopez, R.; Butcher, S. The Embl-Ebi Job Dispatcher Sequence Analysis Tools Framework in 2024. *Nucleic Acids Res.* **2024**, *52*, W521–W525. [CrossRef]
18. Rudra, B.; Gupta, R.S. Phylogenomics Studies and Molecular Markers Reliably Demarcate Genus *Pseudomonas sensu stricto* and Twelve Other *Pseudomonadaceae* Species Clades Representing Novel and Emended Genera. *Front. Microbiol.* **2023**, *14*, 1273665. [CrossRef]
19. Perry, A.; Tambyrajah, W.; Grossmann, J.G.; Lian, L.Y.; Scrutton, N.S. Solution Structure of the Two-Iron Rubredoxin of *Pseudomonas oleovorans* Determined by NMR Spectroscopy and Solution X-Ray Scattering and Interactions with Rubredoxin Reductase. *Biochemistry* **2004**, *43*, 3167–3182. [CrossRef]
20. Sushko, T.; Kavaleuski, A.; Grabovec, I.; Kavaleuskaya, A.; Vakhrameev, D.; Bukhdruker, S.; Marin, E.; Kuzikov, A.; Masamrekh, R.; Shumyantseva, V.; et al. A New Twist of Rubredoxin Function in *M. tuberculosis*. *Bioorganic Chem.* **2021**, *109*, 104721. [CrossRef]
21. Hagelueken, G.; Wiehlmann, L.; Adams, T.M.; Kolmar, H.; Heinz, D.W.; Tümmler, B.; Schubert, W.D. Crystal Structure of the Electron Transfer Complex Rubredoxin Rubredoxin Reductase of *Pseudomonas aeruginosa*. *Proc. Natl. Acad. Sci. USA* **2007**, *104*, 12276–12281. [CrossRef]
22. Lovenberg, W.; Sobel, B.E. Rubredoxin: A New Electron Transfer Protein from *Clostridium pasteurianum*. *Proc. Nat. Acad. Sci. USA* **1965**, *54*, 193–199. [CrossRef]
23. Dauter, Z.; Wilson, K.S.; Sieker, L.C.; Moulis, J.M.; Meyer, J. Zinc- and Iron-Rubredoxins from *Clostridium pasteurianum* at Atomic Resolution: A High-Precision Model of a ZnS_4 Coordination Unit in a Protein. *Proc. Natl. Acad. Sci. USA* **1996**, *93*, 8836–8840. [CrossRef]

https://doi.org/10.3390/biom16050623

24. Bönisch, H.; Schmidt, C.L.; Bianco, P.; Ladenstein, R. Ultrahigh-Resolution Study on *Pyrococcus abyssi* Rubredoxin. I. 0.69 Angstrom X-Ray Structure of Mutant W4l/R5s. *Acta Crystallogr. Sect. D-Biol. Crystallogr.* **2005**, *61*, 990–1004. [CrossRef]
25. Schober, I.; Koblitz, J.; Carbasse, J.S.; Ebeling, C.; Schmidt, M.L.; Podstawka, A.; Gupta, R.; Ilangovan, V.; Chamanara, J.; Overmann, J. BacDive in 2025: The Core Database for Prokaryotic Strain Data. *Nucleic Acids Res.* **2025**, *53*, D748–D756. [CrossRef]
26. Opt, Codon. "Codon Optimization". Available online: https://www.idtdna.com/pages/tools/codon-optimization-tool (accessed on 3 February 2024).
27. Pape, T.; Schneider, T.R. HKL2MAP: A Graphical User Interface for Macromolecular Phasing with Shelx Programs. *J. Appl. Crystallogr.* **2004**, *37*, 843–844. [CrossRef]
28. Casanas, A.; Warshamanage, R.; Finke, A.D.; Panepucci, E.; Olieric, V.; Noll, A.; Tampe, R.; Brandstetter, S.; Forster, A.; Mueller, M.; et al. Eiger Detector: Application in Macromolecular Crystallography. *Acta Crystallogr. Sect. D* **2016**, *72*, 1036–1048. [CrossRef]
29. Kabsch, W. Xds. *Acta Crystallogr. Sect. D* **2010**, *66*, 125–132. [CrossRef] [PubMed]
30. Evans, P. An Introduction to Stereochemical Restraints. *Acta Crystallogr. Sect. D* **2007**, *63*, 58–61. [CrossRef] [PubMed]
31. Evans, P.R.; Murshudov, G.N. How Good Are My Data and What Is the Resolution? *Acta Crystallogr. Sect. D* **2013**, *69*, 1204–1214. [CrossRef]
32. Collaborative Computational Project. The Ccp4 Suite: Programs for Protein Crystallography. *Acta Crystallogr. Sect. D* **1994**, *50*, 760–763. [CrossRef]
33. Vonrhein, C.; Flensburg, C.; Keller, P.; Sharff, A.; Smart, O.; Paciorek, W.; Womack, T.; Bricogne, G. Data Processing and Analysis with the Autoproc Toolbox. *Acta Crystallogr. Sect. D* **2011**, *67*, 293–302. [CrossRef]
34. Altschul, S.F.; Gish, W.; Miller, W.; Myers, E.W.; Lipman, D.J. Basic Local Alignment Search Tool. *J. Mol. Biol.* **1990**, *215*, 403–410. [CrossRef]
35. Duvaud, S.; Gabella, C.; Lisacek, F.; Stockinger, H.; Ioannidis, V.; Durinx, C. Expasy, the Swiss Bioinformatics Resource Portal, as Designed by Its Users. *Nucleic Acids Res.* **2021**, *49*, W216–W227. [CrossRef]
36. Gasteiger, E.; Hoogland, C.; Gattiker, A.; Duvaud, S.; Wilkins, M.R.; Appel, R.D.; Bairoch, A. (Eds.) *Protein Identification and Analysis Tools on the Expasy Server*; Humana Press: Totowa, NJ, USA, 2005; pp. 571–607.
37. Barthels, F.; Schirmeister, T.; Kersten, C. BANΔIT: B'-Factor Analysis for Drug Design and Structural Biology. *Mol. Inform.* **2021**, *40*, 2000144. [CrossRef]
38. Sacquin-Mora, S. Motions and Mechanics: Investigating Conformational Transitions in Multi-Domain Proteins with Coarse-Grain Simulations. *Mol. Simul.* **2014**, *40*, 229–236. [CrossRef]
39. Chen, C.R.; Makhatadze, G.I. Proteinvolume: Calculating Molecular Van Der Waals and Void Volumes in Proteins. *BMC Bioinform.* **2015**, *16*, 101. [CrossRef]
40. DeLano, W.L. *The Pymol Molecular Graphics System*; DeLano Scientific: San Carlos, CA, USA, 2002; Available online: http://www.pymol.org (accessed on 12 March 2026).
41. Ferruz, N.; Schmidt, S.; Höcker, B. Proteintools: A Toolkit to Analyze Protein Structures. *Nucleic Acids Res.* **2021**, *49*, W559–W566. [CrossRef] [PubMed]
42. Vymětal, J.; Jakubec, D.; Galgonek, J.; Vondrášek, J. Amino Acid Interactions (Intaa) Web Server V2.0: A Single Service for Computation of Energetics and Conservation in Biomolecular 3D Structures. *Nucleic Acids Res.* **2021**, *49*, W15–W20. [CrossRef] [PubMed]
43. Blake, P.R.; Park, J.B.; Bryant, F.O.; Aono, S.; Magnuson, J.K.; Eccleston, E.; Howard, J.B.; Summers, M.F.; Adams, M.W. Determinants of Protein Hyperthermostability: Purification and Amino Acid Sequence of Rubredoxin from the Hyperthermophilic Archaebacterium *Pyrococcus furiosus* and Secondary Structure of the Zinc Adduct by Nmr. *Biochemistry* **1991**, *30*, 10885–10895. [CrossRef] [PubMed]
44. Jung, D.H.; Kang, N.S.; Jhon, M.S. Site-Directed Mutation Study on Hyperthermostability of Rubredoxin from *Pyrococcus furiosus* Using Molecular Dynamics Simulations in Solution. *J. Phys. Chem. A* **1997**, *101*, 466–471. [CrossRef]
45. Kurihara, K.; Tanaka, I.; Chatake, T.; Adams, M.W.W.; Jenney, F.E., Jr.; Moiseeva, N.; Bau, R.; Niimura, N. Neutron Crystallographic Study on Rubredoxin from *Pyrococcus furiosus* by Bix-3, a Single-Crystal Diffractometer for Biomacromolecules. *Proc. Nat. Acad. Sci. USA* **2004**, *101*, 11215–11220. [CrossRef] [PubMed]
46. De Maayer, P.; Anderson, D.; Cary, C.; Cowan, D.A. Some Like It Cold: Understanding the Survival Strategies of Psychrophiles. *EMBO Rep.* **2014**, *15*, 508–517. [CrossRef]
47. Feller, G. Protein Stability and Enzyme Activity at Extreme Biological Temperatures. *J. Phys. Condens. Matter* **2010**, *22*, 323101. [CrossRef] [PubMed]
48. Frishman, D.; Argos, P. Knowledge-Based Protein Secondary Structure Assignment. *Proteins Struct. Funct. Bioinform.* **1995**, *23*, 566–579. [CrossRef] [PubMed]
49. Vondrášek, J.; Bendová, L.; Klusák, V.; Hobza, P. Unexpectedly Strong Energy Stabilization inside the Hydrophobic Core of Small Protein Rubredoxin Mediated by Aromatic Residues: Correlated Ab Initio Quantum Chemical Calculations. *J. Am. Chem. Soc.* **2005**, *127*, 2615–2619. [CrossRef]

50. Smalås, A.O.; Leiros, H.K.; Os, V.; Willassen, N.P. Cold Adapted Enzymes. *Biotech. Ann. Rev.* **2000**, *6*, 1–57. [CrossRef]
51. Feller, G. Psychrophilic Enzymes: From Folding to Function and Biotechnology. *Scientifica* **2013**, *2013*, 512840. [CrossRef]

https://doi.org/10.3390/biom16050623

biomolecules

MDPI

Article

A Computational Model for Nme1Cas9 HNH Activation Driven by Dynamic Interface Engineering at Residues S593 and W596

Zhenyu Zhou and Lizhe Zhu *

Warshel Institute for Computational Biology, School of Medicine, The Chinese University of Hong Kong, Shenzhen 518172, China; zhenyuzhou@link.cuhk.edu.cn
* Correspondence: zhulizhe@cuhk.edu.cn

Abstract

Nme1Cas9 is an encouraging genome-editing tool with high fidelity and compactness, but its applications are limited by poor catalytic efficiency compared with SpyCas9. Understanding the dynamic activation mechanism of the HNH nuclease domain is the key to breaking the kinetic bottleneck. Here, we integrated Steered Molecular Dynamics (SMD) with the Traveling-Salesman-based automated Path Searching (TAPS) algorithm to reconstruct the atomic-level activation landscape of the L1-HNH module. The simulations suggest a complex "Lifting-Rearrangement-Sliding" pathway, revealing the critical role of a "Backbone Sliding" conformation; in this step, the HNH domain rotates across the R-loop surface. A thermodynamic analysis using free energy decomposition by MM/PBSA indicates that the intrinsic instability of the wild-type HNH/R-loop interface constitutes the predominant energetic barrier. Hyperactive variants (S593Q/W596K and S593Q/W596R) can overcome this barrier by substantially increasing binding affinity to the R-loop through a "Geometry–Electrostatics Synergism": S593Q improves interfacial proximity, whereas W596K/R acts as an "Electrostatic Anchor." The results of unbiased MD simulations demonstrate that strengthened interfacial interactions effectively promote spontaneous conformational drift toward the activated state. This computational study proposes a novel in silico model for "Dynamic Interface Engineering" in which reinforcing transient interfacial contacts during conformational sliding can be an effective strategy in developing high-efficiency CRISPR-Cas effectors.

Keywords: Nme1Cas9; molecular dynamics; HNH activation

1. Introduction

Clustered regularly interspaced short palindromic repeats (CRISPR) and CRISPR-associated (Cas) systems are advanced prokaryotic adaptive immune systems that are now repurposed as a powerful genome-engineering platform [1–7]. These systems were categorized into two broad classes based on their organization of the effector module: Class 1 systems, which depend on multi-subunit effector complexes, and Class 2 systems, which employ a single, multi-domain effector protein [8,9]. Due to their simplified architecture and programmable operation, Class 2 effectors have provided a new research template for applications ranging from basic biological studies to the clinical therapeutic correction of genetic disorders.

Within this increasingly large Class 2 toolbox, *Neisseria meningitidis* Cas9 (Nme1Cas9) emerges as a promising in vivo candidate. As a compact Type II-C effector consisting of only

 https://doi.org/10.3390/biom16030358

1082 amino acids, Nme1Cas9 possesses a streamlined architecture that allows efficient "all-in-one" packaging into single adeno-associated virus (AAV) vectors, thereby circumventing a major delivery bottleneck in mammalian tissues [10–14]. In addition, Nme1Cas9 possesses extraordinary intrinsic fidelity, with minimal off-target cleavage, and can be finely regulated by natural anti-CRISPR proteins (Acrs), such as AcrIIC3, which serves as a potent off-switch by tethering Cas9 complexes from activating in a particular conformation [15–18]. With these essential therapeutic benefits in mind, Nme1Cas9's general utility is currently limited by its kinetically unfavorable profile, namely its suboptimal intrinsic catalytic activity and cleavage efficiency compared to other widely used orthologs [16,19,20].

The induction mechanism of Nme1Cas9 is a complex process: upon complete R-loop (the tDNA-gRNA heteroduplex) formation, the RuvC and REC2 domains and the L1 linker are subject to large-scale conformational transitions, which induce the HNH domain to escape from its inactive interface with RuvC, make a substantial rotation towards the RNA-DNA heteroduplex, and finally dock near the cleavage site on the target strand [20–22]. Recent comprehensive reviews and computational studies emphasize that such precise conformational checkpoints and allosteric communications are universal master regulators directing the nuclease activity across diverse CRISPR-Cas systems [23,24]. Given the large scale of this conformational excursion of the L1-HNH domain and its underlying structural instability, it is impossible to observe this dynamic process directly using a static structural technique; thus, detailed mechanistic information about the domain rearrangement remains elusive. More importantly, in such a complex enzymatic system, it is commonly observed that the rate-limiting step is governed by the large-scale conformational changes rather than the chemical reaction itself [25]. Thus, a straightforward elucidation of these dynamic processes is a prerequisite for achieving systematic regulation of Cas cleavage efficiency. Indeed, recent advanced molecular dynamics simulations have successfully captured the dynamic interactions and conformational barriers restricting the final activation of the HNH nuclease domain in other Cas9 orthologs, highlighting the power of integrated computational approaches [26]. In particular, pioneering computational studies by Giulia Palermo and co-workers have profoundly reshaped our understanding of Cas9 dynamics. Their extensive molecular dynamics simulations have revealed the striking structural plasticity of the HNH domain and elucidated the long-range allosteric communication networks that govern its activation in SpyCas9 Palermo's work [27,28] demonstrated that HNH activation is not merely an isolated rigid-body swing, but rather a highly correlated process intricately coupled with the motions of the REC lobe and the non-target DNA strand [29]. These milestone studies underscore the necessity of treating Cas9 as a dynamic ensemble and inspire the application of enhanced sampling techniques to capture elusive conformational intermediates.

Earlier experiments by Sun et al. [20] have suggested that stabilizing the HNH domain in its activated state by strengthening HNH-R-loop interactions is a reasonable strategy to enhance the activation efficiency of Nme1Cas9. Following this logic, mutants bearing S593Q/W596K and S593Q/W596R mutations were previously created and rigorously validated through in vitro DNA cleavage assays by Sun et al., resulting in significantly improved catalytic performance that rivals SpyCas9. Here, however, we show that the functional role of residues S593 and W596 extends beyond stabilizing the final active conformation. Using Steered Molecular Dynamics (ABMD from plumed and enforced rotation from Gromacs) [30–32] Molecular Dynamics (MD)-based Traveling Salesman-Based Automated Path Searching (TAPS) [33–36], we reconstructed the dynamic trajectory of the L1-HNH module after complete R-loop pairing. We characterized the sequence of rising, rearrangement, and sliding motions leading to the activated conformation. In the initial lifting process, the HNH domain exits from its inhibitory interface with the RuvC domain. Following L1-HNH reconfiguration, the HNH domain re-approaches the R-loop and, steered by electrostatic guidance residues, rotationally slides along the R-loop surface to finally dock into the catalytic conformation. Importantly, we find that residues S593 and W596 interact with the phosphate backbone of the R-loop as early as this rotation-sliding step. Confirming this observation through extensive unbiased MD experiments and MM-PBSA [37] free energy calculations, we show that these mutations substantially increase the binding affinity to the R-loop during this dynamical transition, facilitating the sliding mechanism of the L1-HNH module and thermodynamically increasing the likelihood of the enzyme to navigate the energy landscape to the active state.

In this paper, we present an all-atomistic simulation of the L1-HNH activation process after complete R-loop pairing and clarify the exact mechanistic functions of residues S593 and W596 and their variants S593Q/W596K and S593Q/W596R in the process of the key rotational sliding conformational transition, by identifying that a stronger binding affinity at the L1-HNH/R-loop interface during sliding favors the transition of the global structure into the activated state, we find a direct causal relation between the stabilization of the intermediate state and the enhancement of the activation efficiency. Our work also demonstrates that the rational design of complex nucleases controlled by large-scale conformational dynamics need not be restricted to static structural templates, but can be efficiently guided by mechanistic understanding derived from dynamical molecular paths.

2. Materials and Methods

2.1. System Construction and Force Field Parameters

The initial models for Nme1Cas9 were constructed based on high-resolution crystal structures. The seed-paired ternary complex (PDB ID: 6KC7) served as the initial state, and the catalytically poised complex (PDB ID: 6JDV) was employed as the final state. Missing residues and disordered loops were modeled using MODELLER (Version 10.5, University of California San Francisco, San Francisco, CA 94143, USA) [38] to ensure structural continuity. Mutant systems (S593Q/W596R and S593Q/W596K) were generated via the PyMOL (Version 3.1.6.1, Schrödinger, LLC, New York, NY, USA) [39] mutagenesis wizard.

Each protein-nucleic acid complex was solvated in a cubic box of TIP3P [40] water molecules with a 10 Å buffer. The systems were neutralized and further ionized with KCl and $MgCl_2$ to achieve final concentrations of 100 mM KCl and 10 mM $MgCl_2$.

2.2. Molecular Dynamics (MD) Simulations

All MD simulations were performed using GROMACS 2019.4 (University of Groningen, Groningen, The Netherlands) with the Amber14SB-OL15 [41,42] force field to describe molecular interactions. Energy minimization was conducted 10,000 steps of steepest descent followed by the conjugate gradient algorithm to remove steric clashes. After minimization, the system was equilibrated in two stages: NVT ensemble at 300 K for 1 ns to equilibrate the solvent and NPT ensemble for 1 ns at 1 atm, controlled using the Berendsen barostat algorithm.

MD simulations used a 1 fs time step with periodic boundary conditions (PBC). Long-range electrostatic interactions were treated using the Particle Mesh Ewald (PME) method, while short-range electrostatics and van der Waals interactions were handled with a 10 Å cutoff. The LINCS algorithm was applied to constrain all bonds.

2.3. Initial Path Generation

To ensure structural stability and accuracy, the initial transition path was generated using a reverse pulling strategy. The starting and ending conformations for this procedure were derived from 50 ns conventional MD simulations initiated from the catalytically

https://doi.org/10.3390/biom16030358

poised state (PDB ID: 6JDV) and the seed-paired complex (PDB ID: 6KC7), respectively. Given that the HNH domain and L1 linker are fully resolved in the 6JDV structure while other regions in 6KC7 are partially incomplete, the activation trajectory was sampled by driving the system from the 6JDV activated state back toward the 6KC7 configuration.

The enhanced sampling combined Enforced Rotation in GROMACS and ABMD in PLUMED (version 2.5.3, SISSA, Trieste, Italy). The Enforced Rotation module was utilized to drive the HNH-L1 domain away from the catalytic site with an isotropic rotation rate of 0.045 deg/ps and a force constant of 500.0 kJ/(mol·nm^2) over a 2 ns simulation. The rotation vector was set to (−0.608, −1.858, −0.883) with a pivot point at (3.3494, 6.8489, 4.2994). To further refine the trajectory, a 2ns ABMD simulation was implemented with a high force constant (KAPPA) of 100,000 kJ/(mol·nm^2). The RMSD of all heavy atoms in the L1 linker and HNH domain served as the collective variable, with alignment performed on the Ca atoms of helical regions in the RuvC, REC1, REC2, and WED domains.

The preliminary trajectory for subsequent path optimization was assembled by concatenating the 2 ns ABMD and 2 ns enforced rotation segments. This integrated path effectively captures the conformational transition between the two functional states and serves as the baseline input for the TAPS path optimization protocol.

2.4. Path Optimization

The TAPS (Traveling-Salesman-Based Automated Path Searching) method was utilized to refine the initial transition path into the low free energy path (LFEP). The theoretical background and methodological details of TAPS have been previously documented in our published work. Path optimization was performed using a custom Python (version 3.5, Python Software Foundation, Wilmington, DE, USA) script (https://github.com/liusong299/TAPS, accessed on 12 December 2025). The convergence of the optimized pathway was validated through PCV-$\sqrt{\langle z \rangle}$ analysis, as illustrated in Figure S6.

2.5. Binding Free Energy Calculation

Binding free energies between the HNH domain and the fully paired R-loop (comprising the gRNA and target DNA) were calculated along the optimized MFEP using the MM-PBSA (Molecular Mechanics Poisson-Boltzmann Surface Area) protocol. The calculations were performed for every frame (interval = 1) using the MMPBSA.py module in AmberTools [43]. The Amber14SB force field and OL15 parameters were employed for the protein and nucleic acids, respectively. To account for the highly charged nature of the DNA-containing system, the GB-Neck2 (igb = 8) implicit solvent model was utilized with a salt concentration of 0.15 M. Crucially, the internal dielectric constant (intdiel) was set to 10, a value recommended for protein-nucleic acid complexes to better capture the screening effects of the polarizable environment.

3. Results

3.1. Dynamic Landscape of L1-HNH Activation

We approximated the full-fledged atomic-level dynamic path of Nme1Cas9 going from its inactive R-loop-pairedits its state to active state using the path optimization algorithm TAPS. To ensure that the optimized pathway was not biased by the relatively large force constant used in the initial Steered MD (SMD) guess, we evaluated the convergence of the trajectories across the iterative TAPS optimization process. As depicted in the Multidimensional Scaling (MDS) projection (Figure S7), the initial aggressive pulling pathway (iter000) systematically relaxed and migrated across the conformational landscape. Ultimately, the pathways from the final iterations (e.g., iter130–137) converged into highly identical trajectories. This convergence quantitatively confirms that the non-equilibrium artifacts from the initial SMD were completely eliminated, yielding a thermodynamically robust, intrinsic low free-energy path (LFEP) for subsequent mechanistic analyses. Activation of HNH domain is a highly non-trivial process. By analyzing the changes in the smallest distance between the important mutation sites (S593/W596) and the R-loop backbone in combination with the spatial displacement and rotation properties of the HNH domain alone the PCV-S [44], we segmented the HNH allosteric activation process, which is a complex L1-HNH allosteric activation, into three kinetic phases (Phase A–C) (Figure 1A).

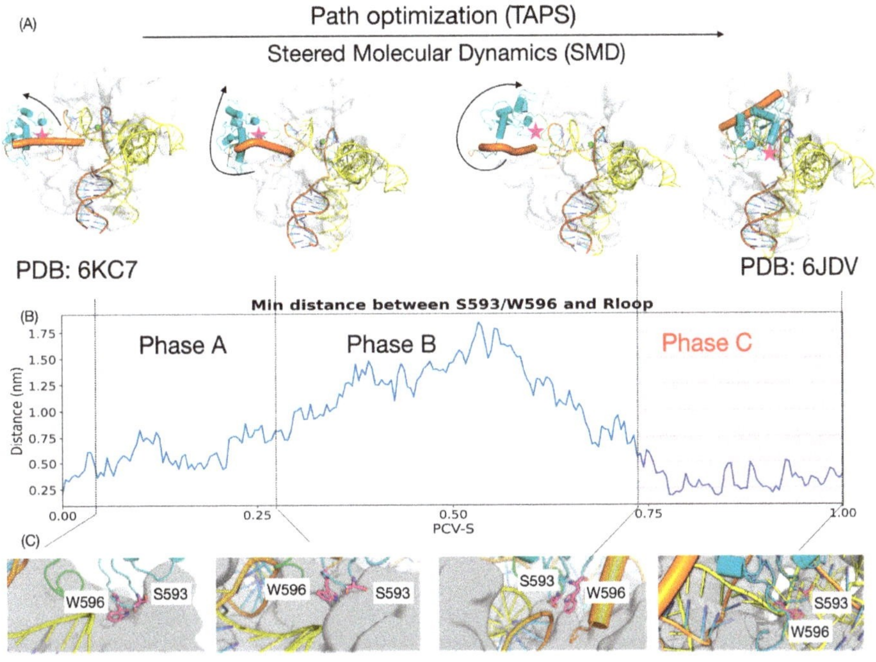

Figure 1. Dynamic activation landscape of the Nme1Cas9 HNH domain revealed by TAPS path optimization. (**A**) Key structural snapshots along the optimized activation pathway. The reaction path was initially generated using Steered Molecular Dynamics (SMD)—specifically combining PLUMED's ABMD and GROMACS's enforced rotation module—and subsequently refined via TAPS path optimization on path collective S (PCV-S) [44]. The sequence illustrates the transition from the inactive state (based on PDB ID: 6KC7) to the final activated state (based on PDB ID: 6JDV), involving domain lifting, L1-HNH rearrangement, and rotation along the R-loop surface. The HNH domain is colored in cyan, the L1 linker in orange, and the sgRNA in yellow. The target DNA strand is depicted with an orange backbone and blue bases. The spatial locations of critical residues S593 and W596 are marked with magenta stars. (**B**) Minimum distance between the mutation sites (S593/W596) and the R-loop backbone along the optimized trajectory. (**C**) Detailed structural close-ups of the local environment around S593 and W596 corresponding to the states shown in figure. Residues S593 and W596 are rendered as magenta sticks, and the L2 loop is highlighted in green. The color scheme for the L1 linker (orange), HNH domain (cyan), and nucleic acids (yellow/orange/blue) remains consistent with (**A**).

3.1.1. Phase A: Domain Lifting and Steric Release

In the initial phase of activation, the HNH domain undergoes a "lifting" motion. During this process, the HNH gradually lifts and slips from its initial contact interface with RuvC domains in the inactive state. This directional displacement is of significant importance, as it effectively eliminates the steric hindrance around the HNH domain and produces the conformational room needed for subsequent large-scale rotational movement of L1-HNH. In the meantime, the distance between S593/W596 and the R-loop gradually expands from 2 Å, which means that the weak interaction between them is very unstable, making L1-HNH in a state of high freedom (Figure 1A–C).

3.1.2. Phase B: Conformational Rearrangement and L1 Restructuring

As the process proceeds, the L1-HNH module proceeds into the next rearrangement phase. We found that the L1-HNH domain begins to exhibit small-scale rotational changes in the process of finding the correct binding orientation. In this process, the minimum distance between S593/W596 and the R-loop exhibits a non-monotonic "expansion-contraction" behavior. It is worth highlighting that such process is accompanied by significant changes in the secondary structure of L1 linker—the L1 takes a special angle of bending and α-Helix folding/bending changes. Such a rigidification of the L1 alpha-helix structure not only limits the random vibration of HNH but also functions as a coiled spring that directs the HNH domain to approach and rotate accurately toward the target R-loop (Figure 1A–C).

3.1.3. Phase C: Electrostatic Sliding and Subsequent Docking (The Decisive "Backbone Sliding" Stage)

During the final, most decisive stage of activation, the HNH domain reaches the completion of large-scale rotation. In contrast to the "lifting" and detachment of Phase A, this stage involves the re-establishment of intimate interfacial contacts between the L1-HNH module and the R-loop. In this phase, the distance between S593/W596 and the R-loop is reduced and stabilized, oscillating within a narrow range of 2–5 Å. This suggests that the L1-HNH domain does not merely "jump" to the active site but instead follows a sophisticated "sliding-rotation" mechanism. Under the electrostatic guidance of surface residues (including S593 and W596), the domain slides along the heteroduplex backbone and gradually optimizes its orientation to overcome the final energetic barrier. This intimate "Backbone Sliding" process is pivotal for locking the HNH domain into the catalytic conformation and accurately orienting its active site toward the scissile phosphodiester bond. As this final phase determines the specificity and cleavage efficiency of the enzyme, the dynamic processes and energy characteristics within this Phase C represent the primary focus of the further mutational and thermodynamic studies.

3.2. Energetic Profile and Critical Metastable Intermediate Analysis During Activation of the HNH Domain

To further explain the thermodynamic driving force for improving the activation efficiency of HNH domain mediated by S593Q and W596R/K mutations, we estimated the time-dependent binding free energy (ΔG_{total}) between HNH domain and the R-loop complex along the TAPS-optimized dynamic route by the MM/PBSA method. From the energy profile (Figure 2A), a key dynamic property of this process can be seen, in the track of the HNH domain from the inactive state to the active state (activation progress). In phase A, the binding energy between L1-HNH and the R-loop gradually rises near 0. In this process, the contact between HNH and the R-loop constantly decreases. In the subsequent phase B, when L1-assisted rotation of L1-HNH occurs, because HNH is very flexible at this stage, it is far from contact with the R-loop and surrounding protein domains, and the binding energy still stays near 0. In phase C, when HNH recontacts the R-loop and starts the large-angle sliding process of L1-HNH on the R-loop surface, the binding energy decreases as contact between L1-HNH and the R-loop increases, to roughly −80 kcal/mol. In phase C, there is a local minimum in the binding energy. We have defined that as the key metastable intermediate "State S". In the critical State S stage, the binding energy advantage of the mutants is significantly enhanced, and the binding energy of S593Q/W596R and S593Q/W596K is roughly 2 kcal/mol lower than WT (Figure 2B). The

https://doi.org/10.3390/biom16030358

residue energy decomposition of the whole activation process ($\Delta\Delta G = \Delta G_{mut} - \Delta G_{WT}$) also showed that this energy reduction can specifically be ascribed to the contribution of positions 593 and 596 to the binding energy after mutation (Figures 2C, S1 and S2). To understand the structural basis of this energy discrepancy, we extracted representative conformations of State S for interface analysis. In State S, the loop region of the HNH domain (sites 593/596) is spatially very close to the DNA phosphate backbone of the R-loop, which is a critical window in establishing intermolecular interactions. However, in the WT system, the side chain of S593 is too short and, because of the geometric constraints, it cannot reach the DNA backbone to a form good interaction with it; at the same time, although W596 had huge side chain volume, the electrostatic attraction between its indole ring's N-H groups with the negatively charged phosphate backbone is very weak and unstable: hence, it is a very poor binding in the WT system in this state (Figure 2D).

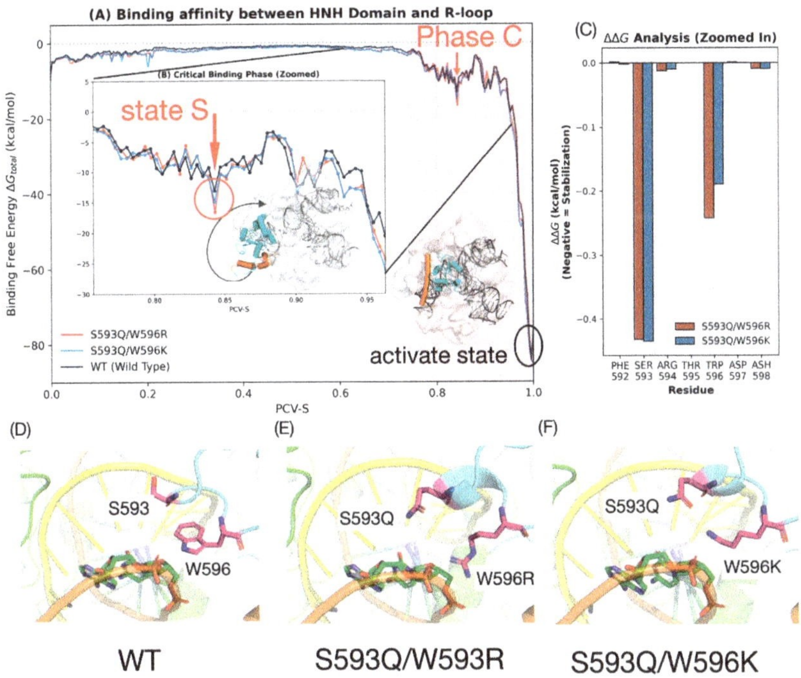

Figure 2. Thermodynamic characterization of the HNH activation pathway and stabilization of the intermediate State S. (**A**) Binding free energy (ΔG_{total}) profiles between the HNH domain and the R-loop along the TAPS-optimized activation trajectory PCV-S. The Wild-type (WT) is shown in black, the S593Q/W596K variant in blue, and the S593Q/W596R variant in red. The structural inset displays the final Activated State, with the HNH domain colored in cyan and the L1 linker in orange. (**B**) A zoomed-in view of the critical binding phase (nested within (**A**)), highlighting the local energy minimum identified as "State S." The structural inset illustrates the conformation of the L1-HNH module at this intermediate state (HNH in cyan, L1 in orange). (**C**) Per-residue binding free energy difference ($\Delta\Delta G$) analysis for the mutation sites and adjacent residues. Values are calculated as $\Delta\Delta G = \Delta G_{mut} - \Delta G_{WT}$ where negative values indicate enhanced stabilization relative to the WT. Blue bars represent S593Q/W596K, and red bars represent S593Q/W596R. (**D**–**F**) Detailed structural comparison of the binding interface at State S for (**D**) WT, (**E**) S593Q/W596R, and (**F**) S593Q/W596K. Residues 593 and 596 are rendered as magenta sticks. The R-loop complex is depicted with an orange backbone and green bases for DNA, and yellow for sgRNA. Note the closer proximity and favorable orientation of the mutant residues toward the DNA backbone compared to the WT.

https://doi.org/10.3390/biom16030358

After the introduction of the mutation, the interaction pattern at this interface changed qualitatively. The S593Q mutation appended a longer side chain to glutamine (Gln), effectively bridging the interface and pulling the residue much closer to the DNA backbone, thereby forming electrostatic or hydrogen-bond networks. In the case of site 596, the substitutions of W596R/K added a strong positive charge. Compared to the neutral tryptophan, the positively charged lysine (Lys) and arginine (Arg) residues produced extreme electrostatic attractions with the negatively charged DNA backbone, like an "electrostatic anchor" for the HNH domain anchoring R-loop. In particular, the W596R mutation, with the complex guanidino backbone from multiple angles, thereby significantly reducing the energy barrier to the conformational transition and stabilizing the active-state conformation (Figure 2E,F).

To rigorously verify that this identified State S is a genuine metastable intermediate rather than a transient artifact of the enhanced sampling process, we subsequently subjected these structures to long-timescale unbiased MD simulations (500 ns $\times$ 2), as detailed in the following Section 3.3.

3.3. Spontaneous Conformational Variant Drift Towards Activated State

To confirm the dynamic characteristics of the variant state "State S" over a long period of simulation time and to explore whether the mutation provides the complex with an intrinsic driving force for the evolutionary transition towards the final activated state, we deduced representative conformations of State S along the TAPS pathway. We conducted unbiased molecular dynamics simulations (500 ns $\times$ 2 replicas) for both WT and the two mutant systems. Although State S is still far from the fully activated state in terms of space conformation (requiring considerable rotation of L1-HNH) compared with the fully activated state defined by the crystal structure, the RMSD analysis of the HNH domain against the activated state (Figure S5) demonstrated dramatic differences. We selected all heavy atoms of L1-HNH to calculate the RMSD. Results showed that the RMSD of the WT system remained high throughout the simulation; that is, the system was typically maintained in this intermediate state, whereas, with no external bias in the forcing terms, the RMSD of both S593Q/W596R and S593Q/W596K systems exhibited a spontaneous, slow-decreasing trend. This conformational drift trend was intuitively confirmed in the multidimensional scaling (MDS) projection of the L1-HNH heavy atoms. The conformational ensemble of the mutants was significantly closer to the reference point representing the activated state (red asterisk) than that of WT (Figure 3B).

To quantify the spatial proximity effect, we examined the dynamic trend of the change in the center-of-mass (COM) distance between the center of the L1-HNH domain and the center of the R-loop. As the simulation time increased, the center distance of the mutant systems showed a dynamic decrease, gradually moving towards the compact activated state conformation, reaching 2.6 nm at the end of the trajectory, whereas that of WT fluctuated around 3.0 nm, demonstrating no obvious directional migration (Figure 3A). Furthermore, the probability density distribution of the center-of-mass distance (Figure 3A) helps quantify this difference. In addition, the WT shows a single peak (~3.0 nm) in its distribution. In contrast, the two mutants show a bimodal distribution, with a substantial subpopulation of formations at a shorter distance (~2.7 nm). These suggest that the contribution via interaction initiated by the mutation effectively reduces the conformational energy barrier, enabling the HNH domain to explore and lock into a spatial position closer to the activated state.

https://doi.org/10.3390/biom16030358

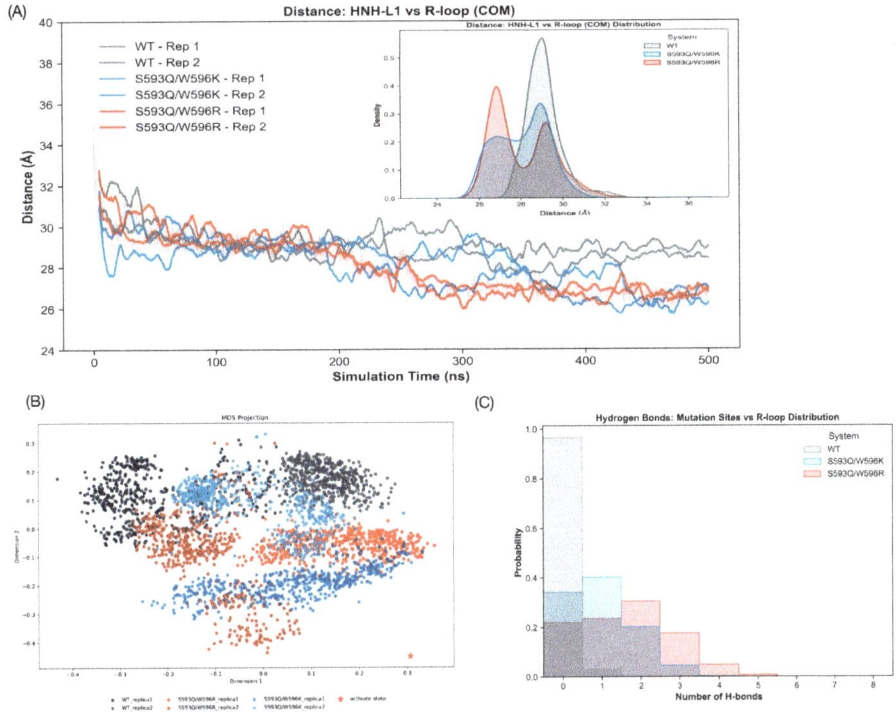

Figure 3. Spontaneous evolution toward the activated state in unbiased MD simulations. (**A**) Time evolution (**left**) and probability distribution (**right**) of the L1-HNH Center-of-Mass (COM) distance between State S and final activated State. Mutant variants show a distinct sub-population at a shorter distance (~2.7 nm). (**B**) 2D MDS projection of the HNH conformational ensemble. The red star indicates the reference activated state. (**C**) Number of hydrogen bonds between residues 593/596 and the R-loop backbone. S593Q/W596R (red) shows the highest occupancy.

The above dynamic tendency and its underlying molecular mechanism are further illustrated in the hydrogen bond analysis and secondary structure stability of the L1 linker (Figures 3C, S3 and S4). The number of hydrogen bonds between the 593/596 position and the R-loop is almost zero in the WT system; thus, the indole ring N-H group of 596 has difficulty maintaining a stable contact with the DNA backbone in the dynamic environment. In contrast, the mutant systems form a stable hydrogen-bonded topology. Particularly, the S593Q/W596R system constructs many more hydrogen bonds than the S593Q/W596K system. When merged with the above-mentioned MMPBSA energy analysis, this result further confirms that the guanidinium head of the arginine (Arg) side chain is of superior geometrical arrangement and has a multi-directional capacity of electrostatic interaction with lysine (Lys), in a stronger manner, can easily grasp the DNA backbone when sliding around HNH, driving conformational adjustments. Structurally, this transition is facilitated by the rigidification of the L1 linker (Figure S3). While WT trajectories exhibit largely disordered coil structures reflecting high intrinsic flexibility, the mutants maintain a continuous ∂-helical conformation. This implies that the variants transform the L1 linker into a stable mechanical element to facilitate HNH domain activation.

https://doi.org/10.3390/biom16030358

4. Discussion

Nme1Cas9 as a High-Fidelity Editor: Mechanistic Insights and Rational Design

The unique properties of Nme1Cas9—especially its small size, high fidelity, and regulatable activity—make this Cas9 variant a promising next-generation gene-editing tool. However, its broader application has been limited by relatively low catalytic efficiency compared with the mature SpyCas9 system [10–22]. Bridging this gap requires a fundamental understanding of the dynamic activation mechanisms that regulate HNH domain transitions. Here, we used Steered Molecular Dynamics (SMD) in conjunction with the TAPS path optimization algorithm to divulge for the first time the complete atomic-level conformational landscape governing the L1-HNH module's transition between the R-loop-bound inactive state and the catalytically competent state.

The simulation results depict a sophisticated "Lifting-Rearrangement-Sliding" activation pathway, in which the HNH domain first gains steric freedom via domain lifting, then undergoes L1-HNH structural rearrangement, and finally undergoes a critical "Backbone sliding" motion along the R-loop surface. To fully contextualize these findings, it is instructive to compare this activation landscape with the well-characterized mechanism of *Streptococcus pyogenes* Cas9 (SpyCas9). In SpyCas9, it is well established that the HNH domain undergoes a massive allosteric rearrangement. As elegantly mapped by Palermo's group through network analysis and MD simulations, the SpyCas9 HNH domain relies on a global, long-range allosteric communication network spanning the REC lobe and the R-loop to reach its catalytically competent state [29,45]. Therefore, the fundamental requirement for large-scale HNH mobility is a conserved feature across Cas9 orthologs. However, the specific "Lifting-Rearrangement-Sliding" trajectory characterized in our study appears to be uniquely tuned to the structural idiosyncrasies of Nme1Cas9. Due to its highly compact architecture and distinct linker compositions compared to SpyCas9, Nme1Cas9 exhibits a more restricted conformational space. Consequently, it relies heavily on specific, transient electrostatic interactions—such as the "Geometry–Electrostatics Synergism" mediated by residues S593 and W596 during the "Backbone Sliding" phase—to precisely navigate its unique conformational energy landscape.

By integrating MM/PBSA free-energy decomposition, we quantified the energetic evolution of the HNH-R-loop interface during the process and identified the crucial roles of residues S593 and W596. We demonstrate that the intrinsic weakness of the interaction between these wild-type residues and the DNA backbone constitutes a substantial energetic barrier to activation. Moreover, we present a robust mechanistic explanation for the enhanced cleavage efficiency of the S593Q/W596K and S593Q/W596R variants, showing how these mutations provide a "Geometry–Electrostatics Synergism". While S593Q finely tunes interfacial distance, W596K/R acts as an "Electrostatic Anchor." The MD simulation results also indicated that these strengthened interactions do not merely stabilize the complex but serve as a kinetic driver, driving a spontaneous conformational drift of the L1-HNH domain towards the activated state. Interestingly, we found that the Arginine variant (W596R) functioned as a more effective "Molecular Gear," leveraging its bidentate hydrogen-bonding ability to promote smooth sliding. At the same time, the mutation-provided rigidity of the L1 linker afforded the required mechanical support for this transition.

This study is not only valuable for providing theoretical insights into the activation mechanisms of CRISPR-Cas systems but also establishes a new computational model for the rational design of high-efficiency Nme1Cas9 variants. According to these findings, future engineering efforts should no longer focus solely on static affinity but also on pursuing a strategy of "Dynamic Interface Engineering": specifically, strengthening the binding interactions between L1-HNH and the R-loop during the sliding process. By reinforcing

https://doi.org/10.3390/biom16030358

this dynamic interface, the entropic penalty of conformational search can be effectively reduced, guiding the HNH domain more smoothly into its catalytic registry. As with any in silico mechanistic model, subsequent studies will extend this design approach to other potential sites and validate the proposed mechanisms through detailed in vitro cleavage assays and in vivo editing experiments.

It should be noted that the absolute binding free energy values calculated via MM/PBSA in this study do not include conformational entropy contributions and may therefore appear exaggerated. However, because our primary focus is on the relative energetic differences (ΔΔG) between structurally similar WT and mutant variants, the entropic contributions are assumed to be largely comparable, making the relative enthalpy-driven trends robust and informative.

5. Conclusions

Based on our in silico simulations, we reconstructed the atomic-level "Lifting-Rearrange-Sliding" activation pathway of Nme1Cas9 using TAPS path optimization, revealing that an essential backbone sliding motion of L1-HNH along the R-loop is mandatory for catalytic competence. Thermodynamic analysis showed that the S593Q/W596K and S594Q/W596R variants strengthen this process via "Geometry–Electrostatics Synergism," with strengthened interfacial contacts promoting the HNH domain towards its active conformations.

We explicitly emphasize that the proposed activation pathways and energetic mechanisms constitute a simulation-based interpretive model. While rigorously supported by thermodynamic calculations, these computational hypotheses warrant further experimental validation via advanced structural and in vivo techniques.

Supplementary Materials: The following supporting information can be downloaded at https://www.mdpi.com/article/10.3390/biom16030358/s1, Figure S1: Energy component decomposition reveals distinct stabilization mechanisms for mutant residues; Figure S2: Ranking of top contributing residues to interfacial binding energetics in Wild-type and variant HNH domains.; Figure S3: Time-evolution of secondary structure stability in the L1 linker; Figure S4: Evolution of interfacial hydrogen bond networks during MD simulations; Figure S5: RMSD evolution of the HNH domain relative to the activated crystal structure (500 ns × 2); Figure S6: Convergence test by calculation for optimized path; Figure S7: Convergence validation of the L1-HNH activation pathway during TAPS optimization; Table S1: Details of ABMD; Table S2: Details of TAPS.

Author Contributions: Conceptualization, Z.Z.; Methodology, Z.Z.; Software, Z.Z.; Validation, Z.Z.; Formal analysis, Z.Z.; Investigation, Z.Z.; Resources, Z.Z.; Data curation, Z.Z.; Writing—original draft, Z.Z.; Writing—review and editing, Z.Z. and L.Z.; Visualization, Z.Z. and L.Z.; Supervision, Z.Z. and L.Z.; Project administration, L.Z.; Funding acquisition, L.Z. All authors have read and agreed to the published version of the manuscript.

Funding: This research was funded by Science, Technology and Innovation Commission of Shenzhen Municipality (Grant No. RCYX20200714114645019), National Science Foundation of China grants (No. 31971179), Warshel Institute for Computational Biology, School of Medicine, The Chinese University of Hong Kong, Shenzhen, Guangdong, China. Department of bioinformatics, School of Medicine, The Chinese University of Hong Kong, Shenzhen, Guangdong, China (No. LGKCSDPT2024001).

Institutional Review Board Statement: Not applicable.

Informed Consent Statement: Not applicable.

Data Availability Statement: The original contributions presented in this study are included in the article/Supplementary Materials. Further inquiries can be directed to the corresponding author.

https://doi.org/10.3390/biom16030358

Acknowledgments: We thank Jinchu Liu and Xi Kun for the fruitful discussion on the manuscript. We acknowledge Xinyu Li for the generous and timely allocation of computational resources, which were essential for this study.

Conflicts of Interest: The authors declare no conflicts of interest.

Abbreviations

The following abbreviations are used in this manuscript:

NmeCas9	*Neisseria meningitidis* Cas9
TAPS	A traveling-salesman based automated path searching method
L1	Linker 1 helix
HNH	Linear dichroism

References

1. Jinek, M.; Chylinski, K.; Fonfara, I.; Hauer, M.; Doudna, J.A.; Charpentier, E. A programmable dual-RNA–guided DNA endonuclease in adaptive bacterial immunity. *Science* **2012**, *337*, 816–821. [CrossRef]
2. O'Connell, M.R.; Oakes, B.L.; Sternberg, S.H.; East-Seletsky, A.; Kaplan, M.; Doudna, J.A. Programmable RNA recognition and cleavage by CRISPR/Cas9. *Nature* **2014**, *516*, 263–266. [CrossRef]
3. Doudna, J.A.; Charpentier, E. The new frontier of genome engineering with CRISPR-Cas9. *Science* **2014**, *346*, 1258096. [CrossRef]
4. Sternberg, S.H.; Redding, S.; Jinek, M.; Greene, E.C.; Doudna, J.A. DNA interrogation by the CRISPR RNA-guided endonuclease Cas9. *Biophys. J.* **2014**, *106*, 695a. [CrossRef]
5. Barrangou, R.; Doudna, J.A. Applications of CRISPR technologies in research and beyond. *Nat. Biotechnol.* **2016**, *34*, 933–941. [CrossRef]
6. Hsu, P.D.; Lander, E.S.; Zhang, F. Development and applications of CRISPR-Cas9 for genome engineering. *Cell* **2014**, *157*, 1262–1278. [CrossRef]
7. Cong, L.; Ran, F.A.; Cox, D.; Lin, S.; Barretto, R.; Habib, N.; Hsu, P.D.; Wu, X.; Jiang, W.; Marraffini, L.A. Multiplex genome engineering using CRISPR/Cas systems. *Science* **2013**, *339*, 819–823. [CrossRef]
8. Makarova, K.S.; Wolf, Y.I.; Iranzo, J.; Shmakov, S.A.; Alkhnbashi, O.S.; Brouns, S.J.; Charpentier, E.; Cheng, D.; Haft, D.H.; Horvath, P. Evolutionary classification of CRISPR–Cas systems: A burst of class 2 and derived variants. *Nat. Rev. Microbiol.* **2020**, *18*, 67–83. [CrossRef] [PubMed]
9. Hille, F.; Richter, H.; Wong, S.P.; Bratovič, M.; Ressel, S.; Charpentier, E. The biology of CRISPR-Cas: Backward and forward. *Cell* **2018**, *172*, 1239–1259. [CrossRef] [PubMed]
10. Hou, Z.; Zhang, Y.; Propson, N.E.; Howden, S.E.; Chu, L.-F.; Sontheimer, E.J.; Thomson, J.A. Efficient genome engineering in human pluripotent stem cells using Cas9 from *Neisseria meningitidis*. *Proc. Natl. Acad. Sci. USA* **2013**, *110*, 15644–15649. [CrossRef] [PubMed]
11. Esvelt, K.M.; Mali, P.; Braff, J.L.; Moosburner, M.; Yaung, S.J.; Church, G.M. Orthogonal Cas9 proteins for RNA-guided gene regulation and editing. *Nat. Methods* **2013**, *10*, 1116–1121. [CrossRef]
12. Zhang, Y.; Heidrich, N.; Ampattu, B.J.; Gunderson, C.W.; Seifert, H.S.; Schoen, C.; Vogel, J.; Sontheimer, E.J. Processing-independent CRISPR RNAs limit natural transformation in *Neisseria meningitidis*. *Mol. Cell* **2013**, *50*, 488–503. [CrossRef] [PubMed]
13. Ibraheim, R.; Tai, P.W.; Mir, A.; Javeed, N.; Wang, J.; Rodríguez, T.C.; Namkung, S.; Nelson, S.; Khokhar, E.S.; Mintzer, E. Self-inactivating, all-in-one AAV vectors for precision Cas9 genome editing via homology-directed repair in vivo. *Nat. Commun.* **2021**, *12*, 6267. [CrossRef] [PubMed]
14. Lee, J.; Mou, H.; Ibraheim, R.; Liang, S.-Q.; Liu, P.; Xue, W.; Sontheimer, E.J. Tissue-restricted genome editing in vivo specified by microRNA-repressible anti-CRISPR proteins. *RNA* **2019**, *25*, 1421–1431. [CrossRef]
15. Kleinstiver, B.P.; Pattanayak, V.; Prew, M.S.; Tsai, S.Q.; Nguyen, N.T.; Zheng, Z.; Joung, J.K. High-fidelity CRISPR–Cas9 nucleases with no detectable genome-wide off-target effects. *Nature* **2016**, *529*, 490–495. [CrossRef] [PubMed]
16. Amrani, N.; Gao, X.D.; Liu, P.; Edraki, A.; Mir, A.; Ibraheim, R.; Gupta, A.; Sasaki, K.E.; Wu, T.; Donohoue, P.D. NmeCas9 is an intrinsically high-fidelity genome-editing platform. *Genome Biol.* **2018**, *19*, 214. [CrossRef]
17. Pawluk, A.; Amrani, N.; Zhang, Y.; Garcia, B.; Hidalgo-Reyes, Y.; Lee, J.; Edraki, A.; Shah, M.; Sontheimer, E.J.; Maxwell, K.L. Naturally occurring off-switches for CRISPR-Cas9. *Cell* **2016**, *167*, 1829–1838.e1829. [CrossRef]
18. Lee, J.; Mir, A.; Edraki, A.; Garcia, B.; Amrani, N.; Lou, H.E.; Gainetdinov, I.; Pawluk, A.; Ibraheim, R.; Gao, X.D. Potent Cas9 inhibition in bacterial and human cells by AcrIIC4 and AcrIIC5 anti-CRISPR proteins. *MBio* **2018**, *9*, e02321-18. [CrossRef]

https://doi.org/10.3390/biom16030358

19. Edraki, A.; Mir, A.; Ibraheim, R.; Gainetdinov, I.; Yoon, Y.; Song, C.-Q.; Cao, Y.; Gallant, J.; Xue, W.; Rivera-Pérez, J.A. A compact, high-accuracy Cas9 with a dinucleotide PAM for in vivo genome editing. *Mol. Cell* **2019**, *73*, 714–726.e714. [CrossRef]
20. Sun, W.; Yang, J.; Cheng, Z.; Amrani, N.; Liu, C.; Wang, K.; Ibraheim, R.; Edraki, A.; Huang, X.; Wang, M. Structures of *Neisseria meningitidis* Cas9 complexes in catalytically poised and anti-CRISPR-inhibited states. *Mol. Cell* **2019**, *76*, 938–952.e935. [CrossRef]
21. Sternberg, S.H.; LaFrance, B.; Kaplan, M.; Doudna, J.A. Conformational control of DNA target cleavage by CRISPR–Cas9. *Nature* **2015**, *527*, 110–113. [CrossRef] [PubMed]
22. Rousseau, B.A.; Hou, Z.; Gramelspacher, M.J.; Zhang, Y. Programmable RNA cleavage and recognition by a natural CRISPR-Cas9 system from *Neisseria meningitidis*. *Mol. Cell* **2018**, *69*, 906–914.e904. [CrossRef]
23. Zhao, S.; Liu, J.; Zuo, Z. Secondary Conformational Checkpoint in CRISPR-Cas9. *J. Chem. Theory Comput.* **2024**, *20*, 3440–3448. [CrossRef] [PubMed]
24. Calvert, R.W.; Knott, G.J. And. . .cut!–how conformational regulation of CRISPR-Cas effectors directs nuclease activity. *Biochem. J.* **2025**, *482*, 1431–1448. [CrossRef]
25. Warshel, A.; Sharma, P.K.; Kato, M.; Xiang, Y.; Liu, H.; Olsson, M.H. Electrostatic basis for enzyme catalysis. *Chem. Rev.* **2006**, *106*, 3210–3235. [CrossRef]
26. Chen, Y.; Li, Y.; Li, P.; Li, X.; Zhao, S.; Zuo, Z. Catching CRISPR-Cas9 in Action. *J. Chem. Theory Comput.* **2025**, *21*, 5023–5036. [CrossRef]
27. Palermo, G.; Miao, Y.; Walker, R.C.; Jinek, M.; McCammon, J.A. Striking plasticity of CRISPR-Cas9 and key role of non-target DNA, as revealed by molecular simulations. *ACS Cent. Sci.* **2016**, *2*, 756–763. [CrossRef]
28. East, K.W.; Newton, J.C.; Morzan, U.N.; Narkhede, Y.B.; Acharya, A.; Skeens, E.; Jogl, G.; Batista, V.S.; Palermo, G.; Lisi, G.P. Allosteric motions of the CRISPR–Cas9 HNH nuclease probed by NMR and molecular dynamics. *J. Am. Chem. Soc.* **2019**, *142*, 1348–1358. [CrossRef]
29. Nierzwicki, L.; East, K.W.; Morzan, U.N.; Arantes, P.R.; Batista, V.S.; Lisi, G.P.; Palermo, G. Enhanced specificity mutations perturb allosteric signaling in CRISPR-Cas9. *eLife* **2021**, *10*, e73601. [CrossRef]
30. Marchi, M.; Ballone, P. Adiabatic bias molecular dynamics: A method to navigate the conformational space of complex molecular systems. *J. Chem. Phys.* **1999**, *110*, 3697–3702. [CrossRef]
31. Tribello, G.A.; Bonomi, M.; Branduardi, D.; Camilloni, C.; Bussi, G. PLUMED 2: New feathers for an old bird. *Comput. Phys. Commun.* **2014**, *185*, 604–613. [CrossRef]
32. Abraham, M.J.; Murtola, T.; Schulz, R.; Páll, S.; Smith, J.C.; Hess, B.; Lindahl, E. GROMACS: High performance molecular simulations through multi-level parallelism from laptops to supercomputers. *SoftwareX* **2015**, *1*, 19–25. [CrossRef]
33. Zhu, L.; Sheong, F.K.; Cao, S.; Liu, S.; Unarta, I.C.; Huang, X. TAPS: A traveling-salesman based automated path searching method for functional conformational changes of biological macromolecules. *J. Chem. Phys.* **2019**, *150*, 124105. [CrossRef]
34. Ti, R.; Pang, B.; Yu, L.; Gan, B.; Ma, W.; Warshel, A.; Ren, R.; Zhu, L. Fine-tuning activation specificity of G-protein-coupled receptors via automated path searching. *Proc. Natl. Acad. Sci. USA* **2024**, *121*, e2317893121. [CrossRef]
35. Li, X.; Liu, Y.; Liu, J.; Ma, W.; Ti, R.; Warshel, A.; Ye, R.D.; Zhu, L. CXC Chemokine Ligand 12 Facilitates Gi Protein Binding to CXC Chemokine Receptor 4 by Stabilizing Packing of the Proline–Isoleucine–Phenylalanine Motif: Insights from Automated Path Searching. *J. Am. Chem. Soc.* **2025**, *147*, 10129–10138. [CrossRef] [PubMed]
36. Xi, K.; Hu, Z.; Wu, Q.; Wei, M.; Qian, R.; Zhu, L. Assessing the performance of traveling-salesman based automated path searching (TAPS) on complex biomolecular systems. *J. Chem. Theory Comput.* **2021**, *17*, 5301–5311. [CrossRef]
37. Valdés-Tresanco, M.S.; Valdés-Tresanco, M.E.; Valiente, P.A.; Moreno, E. gmx_MMPBSA: A new tool to perform end-state free energy calculations with GROMACS. *J. Chem. Theory Comput.* **2021**, *17*, 6281–6291. [CrossRef]
38. Šali, A.; Blundell, T.L. Comparative protein modelling by satisfaction of spatial restraints. *J. Mol. Biol.* **1993**, *234*, 779–815. [CrossRef]
39. DeLano, W.L. Pymol: An open-source molecular graphics tool. *CCP4 Newsl. Protein Crystallogr* **2002**, *40*, 82–92.
40. Mark, P.; Nilsson, L. Structure and dynamics of the TIP3P, SPC, and SPC/E water models at 298 K. *J. Phys. Chem. A* **2001**, *105*, 9954–9960. [CrossRef]
41. Maier, J.A.; Martinez, C.; Kasavajhala, K.; Wickstrom, L.; Hauser, K.E.; Simmerling, C. ff14SB: Improving the accuracy of protein side chain and backbone parameters from ff99SB. *J. Chem. Theory Comput.* **2015**, *11*, 3696–3713. [CrossRef] [PubMed]
42. Zgarbová, M.; Sponer, J.; Otyepka, M.; Cheatham, T.E., III; Galindo-Murillo, R.; Jurecka, P. Refinement of the sugar–phosphate backbone torsion beta for AMBER force fields improves the description of Z-and B-DNA. *J. Chem. Theory Comput.* **2015**, *11*, 5723–5736. [CrossRef] [PubMed]
43. Case, D.A.; Aktulga, H.M.; Belfon, K.; Cerutti, D.S.; Cisneros, G.A.; Cruzeiro, V.W.D.; Forouzesh, N.; Giese, T.J.; Gotz, A.W.; Gohlke, H. AmberTools. *J. Chem. Inf. Model.* **2023**, *63*, 6183–6191. [CrossRef] [PubMed]

https://doi.org/10.3390/biom16030358

44. Hovan, L.; Comitani, F.; Gervasio, F.L. Defining an optimal metric for the path collective variables. *J. Chem. Theory Comput.* **2018**, *15*, 25–32. [CrossRef]
45. Nierzwicki, Ł.; Arantes, P.R.; Saha, A.; Palermo, G. Establishing the allosteric mechanism in CRISPR-Cas9. *Wiley Interdiscip. Rev. Comput. Mol. Sci.* **2021**, *11*, e1503. [CrossRef]

https://doi.org/10.3390/biom16030358

biomolecules

MDPI

Article

Multi-Temperature Crystallography of S-Adenosylmethionine Decarboxylase Observes Dynamic Loop Motions

Jenitha R. Patel, Timothy J. Bonzon, Timothy F. Bakht, Omowumi O. Fagbohun and Jonathan A. Clinger *

Department of Chemistry and Biochemistry, Baylor University, One Bear Place 97348, Waco, TX 76798, USA; jenitha_patel1@baylor.edu (J.R.P.); timothy_bonzon1@baylor.edu (T.J.B.); timothy_bakht1@baylor.edu (T.F.B.); omowumi_fagbohun1@baylor.edu (O.O.F.)
* Correspondence: jonathan_clinger@baylor.edu

Abstract

S-adenosylmethionine decarboxylase (AdoMetDC) is an essential enzyme in the polyamine biosynthesis pathway and plays a key role in the synthesis of the polyamines spermidine and spermine, polycationic alkylamines that are present in millimolar levels in mammalian cells. Polyamines are metabolic molecules that are involved in many fundamental processes, including regulation of protein and nucleic acid synthesis, stabilization of chromatin, differentiation, apoptosis, protection from oxidation, and regulation of ion channels. Multiple oncogenic pathways lead to dysregulation of polyamines, making polyamines a potential biomarker for cancer and polyamine biosynthesis a target for therapeutic intervention. This study uses multi-temperature crystallography to probe the structure and dynamics of AdoMetDC by collecting diffraction data at 100 K, 273 K, and 293 K. Differential loop behavior is observed across the collected datasets, with dramatic residue rearrangements. In the loop containing residues 20–28, the ambient temperature datasets show a large motion relative to the cryo structure. In a second loop containing residues 164–174, previous cryo structures do not report ordered positions. This loop is ordered in our 100 K structure, while assuming different conformations in the 273 K and 293 K data. These results further illustrate the usefulness of ambient data collection for understanding the structure and dynamics of proteins, especially in loop regions which are less restrained than protein cores.

Keywords: S-adenosylmethionine decarboxylase; polyamine; protein dynamics; protein crystallography

1. Introduction

Polyamines are polycations that are found in life across phyla and are important for many cellular growth processes [1,2]. All eukaryotes can synthesize their own polyamines via the polyamine synthesis pathway, and polyamines are essential for cell growth and proliferation in eukaryotes [3]. Balancing polyamine levels is essential, as deviations from their narrow ideal range has severe physiological consequences [4,5]. In humans, the polyamine biosynthetic pathway enzymes consist of ornithine decarboxylase (ODC), S-adenosylmethionine decarboxylase (AdoMetDC), spermidine synthase (SPDS), and spermine synthase (SPS) [6,7]. AdoMetDC converts S-adenosylmethionine (AdoMet) to decarboxylated S-adenosylmethionine (dcAdoMet). dcAdoMet is then used as the aminopropyl donor for SPDS and SPS for their reactions with putrescine and spermidine, respectively [8–10] (Figure 1).

 https://doi.org/10.3390/biom15091274

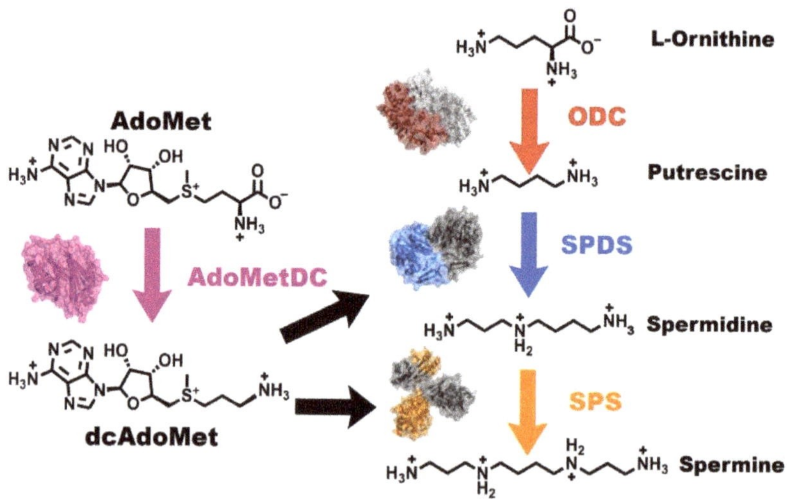

Figure 1. Polyamine biosynthesis pathway with structures of human enzymes. In the co-first steps, S-adenosylmethionine (AdoMet) and L-ornithine are converted to decarboxylated S-adenosylmethionine (dcAdoMet) and putrescine by AdoMetDC (magenta PDB ID 1JEN) and ODC (red/gray PDB ID 1D7K), respectively. Putrescine is then converted to spermidine using dcAdoMet as the aminopropyl donor by SPDS (blue/gray PDB ID 2O07). Spermidine is then converted to spermine using a second dcAdoMet as an aminopropyl donor by SPS (orange/gray PDB ID 3C6K). Gray structures indicate the second chain of homodimer pairs.

Due to the important functions of the polyamine biosynthesis pathway and its association with various disease states, the pathway has long been a target for development of therapeutics. AdoMetDC specifically is rate-limiting for the formation of spermidine and spermine, making it an attractive therapeutic target for modulating polyamine synthesis and cellular levels. A number of inhibitors and drug candidates for AdoMetDC have been proposed and tested in clinical trials [11–14]. The search for new and improved inhibitors of AdoMetDC and other enzymes in the polyamine biosynthesis pathway is on-going, including recent efforts to use virtual screening as well as development of irreversible inhibitors [15–17].

AdoMetDC is a decarboxylase which depends on a pyruvoyl cofactor for its activity, unlike the more common pyridoxal-5′-phosphate dependent enzymes [18]. Other examples of the pyruvoyl-dependent decarboxylation enzymes include aspartate decarboxylase, histidine decarboxylase, and arginine decarboxylase [19–23]. AdoMetDC is expressed as a proenzyme and auto-processes into the active form by an internal serinolysis reaction leading to backbone cleavage to α and β subunits and creation of the pyruvoyl group at the N-terminus of the larger α subunit [24]. This process is activated in humans in the presence of putrescine. Human AdoMetDC (hAdoMetDC) is a dimer in solution according to analytical ultracentrifugation with a 3×10^7 M^{-1} dimerization constant and has been found to cooperatively bind the activating putrescine [25]. However, AdoMetDC concentration is tightly controlled in the living cell by multiple mechanisms, including translational repression, and the cellular half-life of AdoMetDC is as little as one hour [6,26,27]. Previous structural and mutagenic work of this enzyme has been vital to understanding the activation, selectivity, and mechanistic behavior of this enzyme [24,25,28–31].

Even though many structural studies have provided a deep wealth of knowledge about this enzyme, there are still some gaps in our structural understanding of this enzyme.

Previously solved crystal structures contain multiple unmodeled flexible loops, particularly loops containing residues 20–28 (disordered loop 1, DL1), 164–174 (disordered loop 2, DL2), and 292–302 (disordered loop 3, DL3). DL1 and DL2 form a pocket/cleft across from the AdoMet binding pocket. Previous data collections have demonstrated that DL1 becomes more ordered in structures that contain AdoMet mimics (3DZ2, 3DZ3, and 3DZ5) or inhibitors (1I7C, 1I7B, and 1I7M). DL2 has also been reported to be more ordered in a subset of structures containing AdoMet mimics and inhibitors (Figure S1). Most interestingly, a single ambient dataset has been previously collected, containing the methylglyoxal bis(guanylhydrazone) (MGBG) inhibitor [31]. MGBG was the first discovered inhibitor of AdoMetDC, and it binds tightly to the entrance to the AdoMetDC active site [11,31]. This dataset, while lower resolution (2.49 Å), is the only structure to fully model DL2. It also reports more structure in DL1 than many other AdoMetDC depositions, only leaving residues 24–26 unmodeled. Other inhibitor bound states, such as the structure containing the AdoMet mimic 5′-[(3-aminopropyl)methylamino]-5′deoxy-8-methyladenosine (PDB ID 3DZ2), leave 23–26 unmodeled [30]. They also typically leave smaller gaps than the apo structure in DL2, but gaps between residues 166–171 are typical versus the 164–174 gap in apo [29,30]. The variable amounts of order in the datasets generally correspond to more order in structures that contain active site ligands which indicates that DL1 and DL2 are at minimum sensitive to active site perturbation, even though the mechanism of their ordering remains unclear. If the active site structure can perturb the structure of these loops, they could also possibly drive active site conformations as well, which may have implications for inhibitor design by targeting these pockets.

To probe the influence of temperature on the AdoMetDC structure, as the MCBG structure suggests may be possible, we performed multi-temperature crystallography experiments on the apo form of AdoMetDC. Structures previously collected above the glass transition and closer to the physiological temperature in other systems demonstrate structural transitions associated with data collection temperature [32–46]. Previous reported results from multi-temperature work in other systems include allosteric loop opening and closing, as well as alternative ligand binding poses [34,35,38,40,41]. In this study, crystallographic data was collected at 273 K and 293 K as well as 100 K to probe for differences in the structures according to data collection temperature. The new structures of human AdoMetDC reported here have altered loops as a function of data collection temperature, indicating that this enzyme's structure is affected by cryo-cooling and/or cryoprotection. The new structures improve our understanding of the structural ensemble and conformations that are present at closer to physiological temperatures.

2. Materials and Methods

2.1. Protein Expression and Purification

The human S-adenosylmethionine decarboxylase gene *AMD1* was codon-optimized for *E. coli* expression and synthesized (Genscript, Piscataway, NJ, USA). It was cloned into a pET27b vector with an N-terminal 6 × HisTag upstream of a tobacco etch virus protease cleavage site. This vector was transformed into BL21 cells for hAdoMetDC expression. For large-scale expression, the cells were introduced to modified terrific broth media (12 g/L tryptone, 24 g/L yeast extract, 0.4% v/v glycerol, and 0.017 M potassium phosphate monobasic (all Fisher Scientific, Hampton, NH, USA)), containing 50 µg/mL kanamycin from overnight starter cultures. Cultures were shaken at 37 °C and 225 rpm until optical density of 0.6 was reached. Then, 0.5 mM isopropyl β-D-1-thiogalactopyranoside (IPTG) (Gold Bio, St. Louis, MO, USA) was added to induce gene expression. Culture continued at the same settings for 4 h; then, cells were harvested via centrifugation. Cell pellets were frozen overnight at −80 °C. Pelleted cells were resuspended in 40 mL of deionized water, 2.5 mM putrescine, and 0.02% Triton X-100 and lysed via sonication. Cell lysate was clarified by centrifugation, and the supernatant was carried forward for purification via Ni-NTA chromatography (HisTrap HP, Cytivia, Wilmington, DE, USA). The running buffer contained 250 mM NaCl, 100 mM HEPES, and 5 mM imidazole at pH 7.5, and the elution buffer was identical except for increased imidazole concentration to 500 mM. Recombinant hAdoMetDC was found to elute at 190 mM imidazole. Fractions containing hAdoMetDC were further purified via dimethylaminoethyl cellulose (DEAE, BioRad, Hercules, CA, USA) ion exchange chromatography, which was equilibrated with running buffer containing 100 mM HEPES, 1 mM TCEP, 2.5 mM putrescine, and 0.1 mM EDTA, pH 7.5. The elution buffer was identical except for increased NaCl concentration to 2 M. AdoMetDC eluted at 220 mM NaCl.

2.2. Protein Crystallization

hAdoMetDC was concentrated to 5 mg/mL concentration in a crystallization buffer (100 mM HEPES pH 7.5, 2.5 mM putrescine, 1 mM TCEP, and 1 mM 5′-Deoxy-5′-(methylthio)adenosine (MTA)). Sitting drop vapor diffusion trays were prepared with a Formulatrix NT-8 in 2:1 protein to reservoir solution. Crystal growth occurred with conditions of 2–7% PEG 8000, 100 mM Tris-HCl, and pH 8.5, matching the previously reported crystal growth conditions [28–30]. For 100 K data collection, 18% glycerol containing reservoir solution was added to the crystallization drop for cryo-protection to match the previous structures. Previous reports indicate that this percentage of glycerol is vital for 100 K data collection [28–30]. Crystals were then cooled in the cryo-stream on Stanford Synchrotron Radiation Lightsource (SSRL) BL12-1 set to 100 K. For 273 K and 293 K data collection, crystals were not cryoprotected but were covered by RT tubes (MiTeGen, Ithaca, NY, USA) containing reservoir solution to prevent dehydration, and the cryo-stream was set to 273 K or 293 K prior to crystal mounting.

2.3. Data Collection and Processing

Diffraction data were collected at SSRL BL12-1 [47]. Diffraction data was collected in a single sweep from a single crystal for the 100 K and 273 K data collections. The 293 K data collection required three data collections from two crystals to be merged together in order to achieve adequate results. The automated data processing pipeline *xia2* was used to run DIALS (version 3.8) and *AIMLESS* (version 0.7.15) packages for data reduction and merging, respectively [48,49]. PDB ID 1JEN was used as the starting structure for refinement for the 100 K refinement process [28]. The new 100 K structure was used as the search model for molecular replacement in the 273 K and 293 K datasets using *Phaser* (*Phenix* version 1.20) [50]. *Phenix.refine* (*Phenix* version 1.20) was used as the automated refinement pipeline [51]. Coot was used for iterative model building [52]. The 100 K data was deposited in the Protein Data Bank with PDB ID 9P1H, and the 273 K and 293 K data were deposited as 9P7Q and 9PBB, respectively. Raw diffraction images were deposited in the SBGrid databank at data.sbgrid.org. Additional analysis was performed using Ringer and Flipper for sidechain rotamer detection [53]. RoPE was used for comparison of the torsional space with other deposited structures [54]. PASSer was used for prediction of the allosteric sites [55]. Ensemble refinement was performed with the phenix implementation using starting structures, which contained completed loops for DL1, DL2, and DL3 [56]. Structure images were created in PyMol [57]. SBGrid was used as a package manager for maintaining crystallographic software packages [58].

3. Results

3.1. Comparison of PDB ID 9P1H, the New 100 K Apo Structure, and Previously Reported Structures

As a control experiment, a new 100 K crystal dataset of AdoMetDC was collected. The indexing result was P 1 2_1 1 with a unit cell of 73.92, 55.95, 99.17, 90, 110.78, 90, which is very similar to the original apo structure which was deposited as 1JEN [28]. The diffraction data was merged to a maximum resolution of 1.81 Å (see Table S1 for full data processing and refinement statistics of all structures reported in this manuscript). Even though crystals grew better in the presence of 1mM MTA, MTA was not detected in the electron density maps, and instead, Tris was present, as has been previously reported in the apo structure of the processing mutant, and is likely present in 1JEN as well but was not modeled due to limited resolution [24]. Upon analysis of the 100 K dataset, differences were observed between it and 1JEN, as well as other structures containing substituted AdoMet analogues or other potential inhibitors [25,30,31]. The largest deviations from this structure and previous structures are the strong densities present for disordered loop 2 (DL2), which contains residues 164–174 (Figure 2). The most comparable structure, 1JEN, does not contain coordinates for the residues between Phe164 and Gln172. Only one other structure, 1I7C, reports positions for these residues, which has methylglyoxal bis(guanylhydrazone) (MGBG) bound in the AdoMet binding site [31]. However, this loop is in a different orientation than the loop that we observe in our data (Figure 2). This indicates that there are multiple stable loop conformations which can refold preferentially depending on conditions (i.e., ligand, hydration, or temperature). Interestingly, the unique inhibitor MGBG-bound structure was collected at 291K (18 °C), making it different from the rest of the previous work in multiple ways (data collection temperature and unique ligand bound), making it challenging to pinpoint the driver of these structural shifts without additional data.

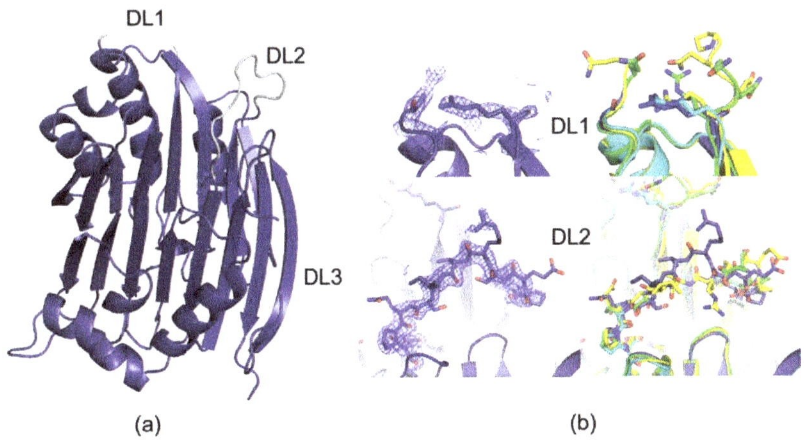

Figure 2. Comparison of new 100 K apo data DL1 and DL2 to previous datasets. (**a**) Overview of the AdoMetDC monomer as observed in 9P1H (indigo). DL1, DL2, and DL3 locations are shown, and DL1 and DL2 are colored in gray. (**b**) Comparison of DL1 and DL2 to previously reported structures. 2Fo-Fc maps contoured at 1.2 RMSD for 9P1H left top (DL1) and left bottom (DL2). Comparison to representative structures 1JEN (apo, 100 K, cyan), 3DZ2 (AdoMet mimic, 100 K, green), and 1I7C (MGBG, 291 K, yellow) for DL1 (right top) and DL2 (right bottom).

The previous structure with the most modeled residues at 100 K for DL2 is 3EP9, a structure without putrescine bound. To prepare this crystal, the putrescine was removed from the protein using perchloric acid prior to crystallization [25]. Other structures, including 1JEN, our structures, and other structures referred to as apo, contain putrescine (Figure S2). 3EP9 has a chain break from Glu166-Gln172. Interestingly, the new 100 K structure has full putrescine occupancy in both monomers, indicating that a more ordered loop does not necessarily correspond to lower putrescine occupancy, as might be surmised by a comparison of previous model structures. In the new structure, the main chain in DL2 is well defined in the 2Fo-Fc map, but the sidechains remain highly flexible and often lack interpretable density (Figure 2). Since the new 100 K form is a dimer in the asymmetric unit, two different monomers may be compared. Density is observed for both; however, the density for the loop in monomer 1 (Chains A and B) is stronger than in monomer 2 (Chains C and D).

DL1 is very similar between the new structure and previous structures, with the most glaring difference being the increased flexibility and disorder of the residues in the new data, relative to other datasets previously reported. We did not assign atomic positions to residues between Arg20 and Gly28 due to poor electron density, and this is consistent with the previously deposited apo structure, 1JEN [28]. However, the other structures, which contain inhibitors and AdoMet analogues, consistently build additional residues for DL1, often up to Pro23 and Gln27 [30]. This consistent increase in buildable positions across so many datasets of similar resolution to the new apo structure indicate that binding AdoMet or a similar molecule on the opposite side of the enzyme confers some structure to DL1. This indicates that DL1 conformation is perturbed by active site occupancy. In the absence of the new data, loss of structure in DL1 could have been assumed to be due to lower quality data in the apo dataset, but now with similar resolution apo data, that appears to not be the case.

3.2. The 273 K and 293 K AdoMetDC Structures Show Loop Fluctuations

Crystals grown under identical conditions to 9P1H, the new 100 K structure, were collected at 273 K and 293 K at SSRL BL12-1 in RT tubes. These crystals indexed in the C 1 2 1 space group that has been reported for many of the inhibitor and AdoMet mimic structures of AdoMetDC. This indicates that the combination of glycerol soaks and cryo-cooling are the source of the symmetry breaking in apo AdoMetDC crystals that results in the P 1 2_1 1 space group. More interestingly, these crystals exhibit differential behavior relative to 9P1H, our new 100 K structure, as well as the other structure collected at ambient temperatures, 1I7C, which contains the inhibitor MGBG. The 273 K and 293 K datasets are best modeled by very similar structures, indicating that there is not a large deviation in structure across this narrow temperature band.

DL2, the loop containing residues 164–174 which is disordered in all previous structures except 1I7C, is again ordered in these datasets. However, DL2 is not in the same orientation as in 9P1H, the new 100 K dataset, but in the same position as 1I7C, the dataset collected at 291K (Figure 3). With the new 273 K and 293 K structures, there are now three total datasets, two apo and one MGBG bound, collected above the glass transition, and all three datasets contain this loop signature. It therefore appears that DL2 is ordered in ambient conditions and repacks during cryo-cooling into a more disordered state. These data imply that DL2's structure is more dependent on data collection temperature than on the presence of an active site ligand. DL1, which contains residues 20–28, is different in both the 273 K and 293 K datasets than previously reported structures. We again do not have the requisite density to build most of the loop as in the other apo structures, 1JEN or 9P1H. However, the tails of the loop behave surprisingly and do not follow the same trend as other 100 K structures or 1I7C. In both the 273 K and 293 K apo datasets, Arg20 is rotated 180 degrees and instead of packing within DL1, it instead forms a 3.1 Å hydrogen bond with Ser171 of DL2 (Figure 3). This twist pulls the loop forward into a more compact structure, and Gln21 can also be modeled confidently. The position of Arg20 would clash with the position of Val169 of DL2 in the new 100 K structure 9P1H, with only 2.1 Å between the Arg20 nitrogen and Val169 carbon. The behavior indicates that during cryo-cooling in the apo structure, DL1 moves to a new position that is favored at cryogenic temperatures. Additionally, Arg20's position in 1I7C being the same as in cryo structures both with and without ligands indicates that the apo structure of DL1 at 100 K is more like the inhibitor-bound state than apo ambient states.

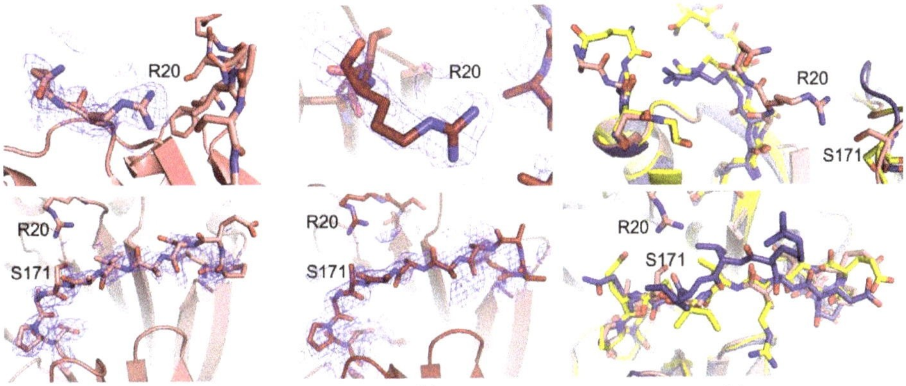

Figure 3. Comparison of DL1 and DL2 from 273 K and 293 K data and 1I7C. (**a**) 2Fo-Fc maps contoured at 1 RMSD for DL1 (top) and DL2 (bottom) of 273 K data. The density only enables modeling of Arg20 and Gln21 (peach, top). It enables a main chain trace through DL2, but the sidechains are poor fits to density in the middle of the loop and thus truncated. (**b**) 2Fo-Fc maps contoured at 1 RMSD for DL1 (top) and DL2 (bottom) of 293 K data. Density only enables modeling of Arg20 (red, top). The density enables a main chain trace through DL2, but the sidechains are poor fits to the density in the middle of the loop and thus truncated. (**c**) Comparison of DL1 and DL2 to previous structures. DL1 loop (273 K, peach) (top panel) is greatly altered compared to previous structures, both apo 100 K (indigo) and 1I7C, which is 291 K and inhibitor-bound (yellow). DL2's main chain forms a structure closely resembling 1I7C (bottom panel). The 293 K data are omitted for clarity.

3.3. Additional Analysis

To assist with placing the new structure data in context with the previous work, we used the Representation of Protein Entities (RoPE) program to compare structures in torsion angle space [54]. Even with a relatively small comparison set, patterns begin to emerge. The closest dataset to the 273 K and 293 K datasets is the 291K MGBG dataset (1I7C), even closer than our new apo 100 K structure, which came from the same protein stock and whose crystals were grown under the same conditions (Figure S3). This was especially true for the larger α chain, which contains DL2. This result indicates that data collection temperature is as important for the torsion angle similarity of AdoMetDC as it is for ligand occupancy. Additionally, in the cluster with the ambient data collections, we also see 3DZ3 and 3DZ5, which are two structures with covalently bound AdoMetDC mimics [30]. 3DZ3 is a Phe223 to alanine mutant with S-Adenosylmethionine methyl ester covalently bound. 3DZ5 is the wild-type enzyme with 5′-[(2-aminooxyethyl)methylamino]-5′-deoxy-8-methyladenosine adducted on the pyruvoyl group. Neither model contains atomic positions for DL1 or DL2, but it is possible that the overall torsional space is less perturbed during cryo-cooling due to the covalent linkages with the AdoMet mimics.

To better visualize the flexible loops and attempt to better understand the possible structural ensemble of the disordered residues, ensemble refinement was completed in *phenix*. The resulting r-free scores were improvements from the single deposited models for the 100 K and 273 K datasets; however, these are considered illustrative only, as clashes increase, as well as the R-work/R-free gap (Table S2). In these refinements, the 100 K structure's DL1 residues are elongated, similarly to previously deposited structures collected at 100 K and containing ligands. Conversely, the 273 K and 293 K refinements show a more condensed DL1 loop, which folds forward more closely over the putrescine binding pocket (Figure 4). DL2 maintains the same general shape as in the deposited single models but is still quite dynamic, with its position moving towards DL1 in the 100 K model relative to the ambient temperature data collections. Disordered loop 3 (DL3), which was not able to be confidently modeled in any of our datasets, is highly mobile in the ensemble refinements. This loop appears greatly frustrated in the crystal structures, and previous structures have also struggled to model this loop.

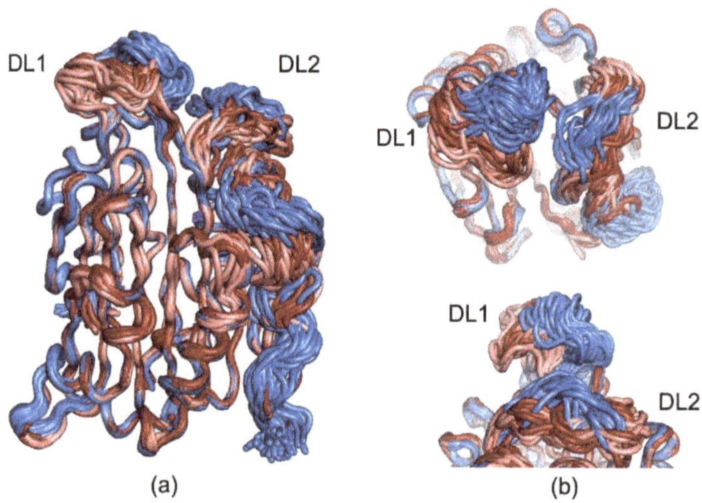

Figure 4. Ensemble refinements of 100 K, 273 K, and 293 K data for dynamic loop visualization. (**a**) Aligned ensemble refinements' structure overview. The 100 K ensemble is shown in blue, the 273 K ensemble shown in peach, and the 293 K ensemble shown in red. DL1 and DL2 are labelled. (**b**) Closer inspection of DL1 and DL2. Top, forward rotation of structure in panel a shows ambient temperature data DL1 leaning forward in the cleft relative to 100 K, which has a more typical orientation for other structures in the PDB. DL2 shifts with DL1, as in 100 K, DL2 rises into a position that would clash with ambient DL1 positions. Bottom, side view of the same structures to better demonstrate how much further forward in the structure DL1 is in the ambient ensembles than at 100 K.

To further investigate whether the DL1 or DL2 loops or other clefts in AdoMetDC are allosteric in nature, we submitted the 100 K structure to the Protein Allosteric Sites Server (PASSer) [55]. PASSer uses three trained machine learning models to predict allosteric sites in proteins and returns ranked scores and probabilities of pockets to be allosteric in nature. PASSer returned three possible allosteric pockets, which included the cleft between DL1 and DL2, as well as the putrescine binding site and the active site (Supplemental Figure S4). These results are also consistent with our observations of different DL1 behavior in the presence and absence of ligands bound when comparing our ambient apo data to previous structures which contained AdoMet mimics or other inhibitors.

Finally, we also used the Ringer/Flipper pipeline to analyze sidechain heterogeneity between the data collected at different temperatures (Figure S5) [33,41,53]. The sidechain analysis was inconclusive, as 273 K had increased heterogeneity relative to 100 K, while 293 K showed decreased heterogeneity. This could be due to the 293 K structure being the least structurally frustrated of the structures; however, it could also be due to differences in data quality between the crystals/datasets or some other confounding effect.

4. Discussion

These results suggest cryo-cooling and/or cryo-protection affects mobile loops in AdoMetDC, causing differential folding above and below the glass transition. The new 100 K structure, 9P1H, reveals a new conformation for DL2, which has not been observed before in other structures. It has moved towards the typical position of DL1 from structures containing inhibitors which have more of DL1 modeled than our structure. DL1 behaves as it does in other apo datasets in 9P1H and the placement of Arg20 is in line with the general trend for other AdoMetDC structures. These loops are therefore clearly flexible and can inhabit a wide range of conformations. In the new data reported in the present work, the structure for loop 164–174 at 100 K is a welcome change and an indication that this loop can form multiple stable orientations. It is unclear why the new 100 K structure has a more ordered DL2 loop than previous structures. The crystal conditions are very similar, with only small variations between salt and PEG conditions, as well as the same crystal space group with very similar unit cell dimensions. The crystal was also cryo-preserved, similar to previous 100 K structures in the same final concentration of glycerol. It is also probably not due to differences in crystal cooling rate or crystal age, as crystals were 1–2 weeks old, as previously reported, and many previous structures were cooled in the cold stream at the beamline or with the home source, as was the crystal in this study [28,30]. Small changes in crystal hydration may play a role, and future studies further perturbing the structural landscape using variable humidity data collection or high pressure cryo-cooling may further uncover perturbations of AdoMetDC's structure [59–62].

Additionally, the altered conformations in 273 K and 293 K indicate that DL1 and DL2 are both affected by cryo-protection and cryo-cooling, causing them to move significantly away from the preferred structures at ambient conditions. Interestingly, the new 273 K and 293 K datasets (9P7Q and 9PBB, respectively) maintain a nearly identical mainchain trace with 1I7C, the only other ambient structure, even though it contains an inhibitor and they do not. However, Arg20 of the DL1 loop is greatly altered compared to previous structures, including 1I7C and 9P1H, the new 100 K structure. These data are highly suggestive of DL2 being temperature-sensitive, with collection temperature contributing to the altered position and order of the loop as well as the active site ligand. However, DL1 may have allosteric behavior that has been previously obscured in 100 K structures, as evidenced by the differences between ambient and cryogenic data in this work in combination with the ambient inhibitor structure from 1I7C. The new conformations observed in this work suggest that prior apo structures' DL1 and DL2 loops shift during cryo-cooling processes, with these shifts hiding the true apo behavior of DL1 and obscuring the structure of DL2. The higher temperature structures are both more compact, with greater interplay between loops 1 and 2, including hydrogen bond formation, which occludes the interior of the protein above the putrescine binding site.

These results contribute to a growing body of literature demonstrating the importance of temperature in the structure and dynamics of proteins. Protein energy landscapes are complex and sometimes unintuitive, whereas their structure can become more ordered at higher temperatures, especially as enzymes reach their temperature of adaptation [32,35,46]. This complex behavior may partially explain the challenges involved with virtual screening and other drug design paradigms that make heavy use of PDB structures, of which the vast majority are collected at cryogenic temperatures. Such downstream efforts may be improved by protein target structural biology derived from physiological or ambient data collections, making the starting structure for the search more like the majority species in the living cell. This problem is not fully addressed by short molecular dynamics simulations used to equilibrate structures prior to docking, as the timescales of even flash cryo-cooling protein crystals allow for motions significantly slower (milliseconds) than typical all-atom molecular dynamics simulations (nanoseconds to microseconds) [63–66].

5. Conclusions

hAdoMetDC, an enzyme required for polyamine biosynthesis and whose activity is tightly controlled by multiple mechanisms within the cell, has long proved to be an elusive drug target to attack metabolic dysfunction in diseases including cancer. This work reports fluctuations in the structure of potentially allosteric loops, including salt bridge formation leading to a more closed conformation than previously observed. These results reinforce the usefulness of multi-temperature data collection for understanding the structure and dynamics of proteins, especially for proteins which contain flexible loop regions and other regions which may be structurally impacted by cryo-cooling and penetrating cryo-protectants. These structures also serve as alternative starting points for downstream experiments that use high-resolution structure information, such as molecular dynamics simulations and virtual screening. Loop fluctuations as detailed here would take long simulation runs to transition from the cryogenic starting points to the ambient positions, and there is no guarantee that they would ever converge on this structure without prior knowledge. Additionally, the behavior of DL1 introduces a second pocket other than the active site that potentially could be used for structure-based drug design efforts.

Supplementary Materials: The following supporting information can be downloaded at: https://www.mdpi.com/article/10.3390/biom15091274/s1, Table S1: Diffraction data processing and refinement statistics for PDB depositions; Figure S1: Location of binding pockets in relation to DL1 and DL2; Figure S2: Putrescine pocket of 100 K, 273 K, and 293 K structures; Figure S3: RoPE analysis of AdoMetDC alpha chain; Table S2: Ensemble refinement statistics from phenix.ensemble_refinement; Figure S4: Protein Allosteric Sites Server (PASSer) results; Figure S5: Ringer/Flipper peak gain/loss analysis of AdoMetDC.

Author Contributions: Conceptualization, J.R.P. and J.A.C.; methodology, J.R.P. and T.F.B.; validation, J.R.P., T.J.B. and O.O.F.; formal analysis, J.R.P., T.J.B., O.O.F. and J.A.C.; investigation, J.R.P., T.F.B. and J.A.C.; resources, J.A.C.; data curation, J.R.P. and J.A.C.; writing—original draft preparation, J.R.P., T.J.B., O.O.F. and J.A.C.; writing—review and editing, J.R.P. and J.A.C.; visualization, J.R.P., T.J.B., O.O.F. and J.A.C.; supervision, J.A.C.; project administration, J.A.C.; funding acquisition, J.R.P. and J.A.C. All authors have read and agreed to the published version of the manuscript.

Funding: This research was funded by a grant from the Cancer Prevention Research Institute of Texas (CPRIT) award RR220081 to J.A.C. This study was supported in part by funds from the Undergraduate Research and Scholarly Achievement Research Grant Program and the Office of Engaged Learning at Baylor University awarded to J.R.P.

Institutional Review Board Statement: Not applicable.

Informed Consent Statement: Not applicable.

Data Availability Statement: The 100 K, 273 K, and 293 K structures were deposited in the PDB with IDs 9P1H, 9P7Q, and 9PBB, respectively, along with reflections used for refinement, scaled and unmerged data, and the map from the final round of refinement. The raw diffraction images from which those reflections were measured were deposited in the sbgrid.org databank with the following DOIs: 9P1H 100 K, https://doi.org/10.15785/SBGRID/1188; 9P7Q 273 K, https://doi.org/10.15785/SBGRID/1189; and 9PBB 293 K, https://doi.org/10.15785/SBGRID/1190.

Acknowledgments: The data were collected at SSRL beamline 12-1. Use of the Stanford Synchrotron Radiation Lightsource, SLAC National Accelerator Laboratory, is supported by the U.S. Department of Energy, Office of Science, Office of Basic Energy Sciences under contract No. DE-AC02-76SF00515. The SSRL Structural Molecular Biology Program is supported by the DOE Office of Biological and Environmental Research and by the National Institutes of Health, National Institute of General Medical Sciences (P30GM133894). The contents of this publication are solely the responsibility of the authors and do not necessarily represent the official views of the NIGMS or NIH. Additional crystal screening experiments in support of this work were conducted at the Cornell High Energy Synchrotron Source beamline 7B2. This work is based on research conducted at the Center for High-Energy X-ray Sciences (CHEXS), which is supported by the National Science Foundation (BIO, ENG, and MPS Directorates) under award DMR-2342336, and the Macromolecular Diffraction at CHESS (MacCHESS) facility, which is supported by award 1-P30-GM124166 from the National Institute of General Medical Sciences and the National Institutes of Health. The authors thank the CMI (Center for Microscopy and Imaging) at Baylor University (Waco, TX) for technical support during microscopy and image analysis. This research was supported in part by the Baylor University Molecular Biosciences Center (Waco, TX).

Conflicts of Interest: The authors declare no conflicts of interest.

Abbreviations

The following abbreviations are used in this manuscript:

ODC	Ornithine decarboxylase
AdoMetDC	S-adenosylmethionine decarboxylase
SPDS	Spermidine synthase
SPS	Spermine synthase
AdoMet	S-adenosylmethionine
dcAdoMet	Decarboxylated S-adenosylmethionine
hAdoMetDC	Human S-adenosylmethionine decarboxylase
DL1	Disordered loop 1
DL2	Disordered loop 2
DL3	Disordered loop 3
AMD1	Human AdoMetDC gene
IPTG	Isopropyl β-D-1-thiogalactopyranoside
MTA	5′-Deoxy-5′-(methylthio)adenosine
MGBG	Methylglyoxal bis(guanylhydrazone)
RoPE	Representation of Protein Entities
PASSer	Protein Allosteric Sites Server

References

1. Michael, A.J. Polyamines in Eukaryotes, Bacteria, and Archaea. *J. Biol. Chem.* **2016**, *291*, 14896–14903. [CrossRef] [PubMed]
2. Michael, A.J. Biosynthesis of Polyamines and Polyamine-Containing Molecules. *Biochem. J.* **2016**, *473*, 2315–2329. [CrossRef] [PubMed]
3. Xuan, M.; Gu, X.; Li, J.; Huang, D.; Xue, C.; He, Y. Polyamines: Their Significance for Maintaining Health and Contributing to Diseases. *Cell Commun. Signal.* **2023**, *21*, 348. [CrossRef]
4. Moinard, C.; Cynober, L.; de Bandt, J.-P. Polyamines: Metabolism and Implications in Human Diseases. *Clin. Nutr.* **2005**, *24*, 184–197. [CrossRef]

5. Bae, D.-H.; Lane, D.J.R.; Jansson, P.J.; Richardson, D.R. The Old and New Biochemistry of Polyamines. *Biochim. Biophys. Acta (BBA)-Gen. Subj.* **2018**, *1862*, 2053–2068. [CrossRef]
6. Pegg, A.E. Mammalian Polyamine Metabolism and Function. *IUBMB Life* **2009**, *61*, 880–894. [CrossRef]
7. Pegg, A.E.; Casero, R.A., Jr. (Eds.) Current Status of the Polyamine Research Field. In *Polyamines: Methods and Protocols*; Humana Press: Totowa, NJ, USA, 2011; pp. 3–35, ISBN 978-1-61779-034-8.
8. Ikeguchi, Y.; Bewley, M.C.; Pegg, A.E. Aminopropyltransferases: Function, Structure and Genetics. *J. Biochem.* **2006**, *139*, 1–9. [CrossRef]
9. Wu, H.; Min, J.; Ikeguchi, Y.; Zeng, H.; Dong, A.; Loppnau, P.; Pegg, A.E.; Plotnikov, A.N. Structure and Mechanism of Spermidine Synthases. *Biochemistry* **2007**, *46*, 8331–8339. [CrossRef] [PubMed]
10. Wu, H.; Min, J.; Zeng, H.; McCloskey, D.E.; Ikeguchi, Y.; Loppnau, P.; Michael, A.J.; Pegg, A.E.; Plotnikov, A.N. Crystal Structure of Human Spermine Synthase: Implications of Substrate Binding and Catalytic Mechanism. *J. Biol. Chem.* **2008**, *283*, 16135–16146. [CrossRef]
11. Williams-Ashman, H.G.; Schenone, A. Methyl Glyoxal Bis(Guanylhydrazone) as a Potent Inhibitor of Mammalian and Yeast S-Adenosylmethionine Decarboxylases. *Biochem. Biophys. Res. Commun.* **1972**, *46*, 288–295. [CrossRef]
12. Regenass, U.; Mett, H.; Stanek, J.; Mueller, M.; Kramer, D.; Porter, C.W. CGP 48664, a New S-Adenosylmethionine Decarboxylase Inhibitor with Broad Spectrum Antiproliferative and Antitumor Activity1. *Cancer Res.* **1994**, *54*, 3210–3217. [PubMed]
13. Millward, M.J.; Joshua, A.; Kefford, R.; Aamdal, S.; Thomson, D.; Hersey, P.; Toner, G.; Lynch, K. Multi-Centre Phase II Trial of the Polyamine Synthesis Inhibitor SAM486A (CGP48664) in Patients with Metastatic Melanoma. *Invest. New Drugs* **2005**, *23*, 253–256. [CrossRef] [PubMed]
14. Siu, L.L.; Rowinsky, E.K.; Hammond, L.A.; Weiss, G.R.; Hidalgo, M.; Clark, G.M.; Moczygemba, J.; Choi, L.; Linnartz, R.; Barbet, N.C.; et al. A Phase I and Pharmacokinetic Study of SAM486A, a Novel Polyamine Biosynthesis Inhibitor, Administered on a Daily-Times-Five Every-Three-Week Schedule in Patients with Advanced Solid Malignancies1. *Clin. Cancer Res.* **2002**, *8*, 2157–2166. [PubMed]
15. Muthukumaran, S.; Sulochana, K.N.; Umashankar, V. Structure Based Design of Inhibitory Peptides Targeting Ornithine Decarboxylase Dimeric Interface and in Vitro Validation in Human Retinoblastoma Y79 Cells. *J. Biomol. Struct. Dyn.* **2021**, *39*, 5261–5275. [CrossRef]
16. Zhou, X.E.; Suino-Powell, K.; Schultz, C.R.; Aleiwi, B.; Brunzelle, J.S.; Lamp, J.; Vega, I.E.; Ellsworth, E.; Bachmann, A.S.; Melcher, K. Structural Basis of Binding and Inhibition of Ornithine Decarboxylase by 1-Amino-Oxy-3-Aminopropane. *Biochem. J.* **2021**, *478*, 4137–4149. [CrossRef]
17. Schultz, C.R.; Aleiwi, B.; Zhou, X.E.; Suino-Powell, K.; Melcher, K.; Almeida, N.M.S.; Wilson, A.K.; Ellsworth, E.L.; Bachmann, A.S. Design, Synthesis, and Biological Activity of Novel Ornithine Decarboxylase (ODC) Inhibitors. *J. Med. Chem.* **2025**, *68*, 5760–5773. [CrossRef]
18. Bale, S.; Ealick, S.E. Structural Biology of S-Adenosylmethionine Decarboxylase. *Amino Acids* **2010**, *38*, 451–460. [CrossRef]
19. Gallagher, T.; Rozwarski, D.A.; Ernst, S.R.; Hackert, M.L. Refined Structure of the Pyruvoyl-Dependent Histidine Decarboxylase from *Lactobacillus* 30a. *J. Mol. Biol.* **1993**, *230*, 516–528. [CrossRef]
20. Albert, A.; Dhanaraj, V.; Genschel, U.; Khan, G.; Ramjee, M.K.; Pulido, R.; Sibanda, B.L.; von Delft, F.; Witty, M.; Blundell, T.L.; et al. Crystal Structure of Aspartate Decarboxylase at 2.2 Å Resolution Provides Evidence for an Ester in Protein Self–Processing. *Nat. Struct. Mol. Biol.* **1998**, *5*, 289–293. [CrossRef]
21. Schmitzberger, F.; Kilkenny, M.L.; Lobley, C.M.C.; Webb, M.E.; Vinkovic, M.; Matak-Vinkovic, D.; Witty, M.; Chirgadze, D.Y.; Smith, A.G.; Abell, C.; et al. Structural Constraints on Protein Self-processing in L-aspartate-α-decarboxylase. *EMBO J.* **2003**, *22*, 6193–6204. [CrossRef]
22. Soriano, E.V.; McCloskey, D.E.; Kinsland, C.; Pegg, A.E.; Ealick, S.E. Structures of the N47A and E109Q Mutant Proteins of Pyruvoyl-Dependent Arginine Decarboxylase from *Methanococcus jannaschii*. *Acta Crystallogr. Sect. D Biol. Crystallogr.* **2008**, *64*, 377–382. [CrossRef]
23. Tolbert, W.D.; Graham, D.E.; White, R.H.; Ealick, S.E. Pyruvoyl-Dependent Arginine Decarboxylase from *Methanococcus jannaschii*: Crystal Structures of the Self-Cleaved and S53A Proenzyme Forms. *Structure* **2003**, *11*, 285–294. [CrossRef]
24. Ekstrom, J.L.; Tolbert, W.D.; Xiong, H.; Pegg, A.E.; Ealick, S.E. Structure of a Human S-Adenosylmethionine Decarboxylase Self-Processing Ester Intermediate and Mechanism of Putrescine Stimulation of Processing As Revealed by the H243A Mutant. *Biochemistry* **2001**, *40*, 9495–9504. [CrossRef]
25. Bale, S.; Lopez, M.M.; Makhatadze, G.I.; Fang, Q.; Pegg, A.E.; Ealick, S.E. Structural Basis for Putrescine Activation of Human S-Adenosylmethionine Decarboxylase. *Biochemistry* **2008**, *47*, 13404–13417. [CrossRef]
26. Miller-Fleming, L.; Olin-Sandoval, V.; Campbell, K.; Ralser, M. Remaining Mysteries of Molecular Biology: The Role of Polyamines in the Cell. *J. Mol. Biol.* **2015**, *427*, 3389–3406. [CrossRef]

27. Yordanova, M.M.; Loughran, G.; Zhdanov, A.V.; Mariotti, M.; Kiniry, S.J.; O'Connor, P.B.F.; Andreev, D.E.; Tzani, I.; Saffert, P.; Michel, A.M.; et al. AMD1 mRNA Employs Ribosome Stalling as a Mechanism for Molecular Memory Formation. *Nature* **2018**, *553*, 356–360. [CrossRef] [PubMed]
28. Ekstrom, J.L.; Mathews, I.I.; Stanley, B.A.; Pegg, A.E.; Ealick, S.E. The Crystal Structure of Human S-Adenosylmethionine Decarboxylase at 2.25 Å Resolution Reveals a Novel Fold. *Structure* **1999**, *7*, 583–595. [CrossRef] [PubMed]
29. Bale, S.; Brooks, W.; Hanes, J.W.; Mahesan, A.M.; Guida, W.C.; Ealick, S.E. Role of the Sulfonium Center in Determining the Ligand Specificity of Human S-Adenosylmethionine Decarboxylase. *Biochemistry* **2009**, *48*, 6423–6430. [CrossRef]
30. McCloskey, D.E.; Bale, S.; Secrist, J.A.I.; Tiwari, A.; Moss, T.H.I.; Valiyaveettil, J.; Brooks, W.H.; Guida, W.C.; Pegg, A.E.; Ealick, S.E. New Insights into the Design of Inhibitors of Human S-Adenosylmethionine Decarboxylase: Studies of Adenine C8 Substitution in Structural Analogues of S-Adenosylmethionine. *J. Med. Chem.* **2009**, *52*, 1388–1407. [CrossRef] [PubMed]
31. Tolbert, W.D.; Ekstrom, J.L.; Mathews, I.I.; Secrist, J.A.; Kapoor, P.; Pegg, A.E.; Ealick, S.E. The Structural Basis for Substrate Specificity and Inhibition of Human S-Adenosylmethionine Decarboxylase. *Biochemistry* **2001**, *40*, 9484–9494. [CrossRef]
32. Keedy, D.A.; van den Bedem, H.; Sivak, D.A.; Petsko, G.A.; Ringe, D.; Wilson, M.A.; Fraser, J.S. Crystal Cryocooling Distorts Conformational Heterogeneity in a Model Michaelis Complex of DHFR. *Structure* **2014**, *22*, 899–910. [CrossRef] [PubMed]
33. Keedy, D.A.; Kenner, L.R.; Warkentin, M.; Woldeyes, R.A.; Hopkins, J.B.; Thompson, M.C.; Brewster, A.S.; Van Benschoten, A.H.; Baxter, E.L.; Uervirojnangkoorn, M.; et al. Mapping the Conformational Landscape of a Dynamic Enzyme by Multitemperature and XFEL Crystallography. *eLife* **2015**, *4*, 07574. [CrossRef] [PubMed]
34. Fischer, M.; Shoichet, B.K.; Fraser, J.S. One Crystal, Two Temperatures: Cryocooling Penalties Alter Ligand Binding to Transient Protein Sites. *ChemBioChem* **2015**, *16*, 1560–1564. [CrossRef]
35. Keedy, D.A.; Hill, Z.B.; Biel, J.T.; Kang, E.; Rettenmaier, T.J.; Brandão-Neto, J.; Pearce, N.M.; von Delft, F.; Wells, J.A.; Fraser, J.S. An Expanded Allosteric Network in PTP1B by Multitemperature Crystallography, Fragment Screening, and Covalent Tethering. *eLife* **2018**, *7*, e36307. [CrossRef]
36. Doukov, T.; Herschlag, D.; Yabukarski, F. Instrumentation and Experimental Procedures for Robust Collection of X-Ray Diffraction Data from Protein Crystals across Physiological Temperatures. *J. Appl. Crystallogr.* **2020**, *53*, 1493–1501. [CrossRef]
37. Fischer, M. Macromolecular Room Temperature Crystallography. *Q. Rev. Biophys.* **2021**, *54*, e1. [CrossRef] [PubMed]
38. Bradford, S.Y.C.; El Khoury, L.; Ge, Y.; Osato, M.; Mobley, D.L.; Fischer, M. Temperature Artifacts in Protein Structures Bias Ligand-Binding Predictions. *Chem. Sci.* **2021**, *12*, 11275–11293. [CrossRef]
39. Ebrahim, A.; Riley, B.T.; Kumaran, D.; Andi, B.; Fuchs, M.R.; McSweeney, S.; Keedy, D.A. The Tem per ature-Dependent Conformational Ensemble of SARS-CoV-2 Main Protease (Mpro). *IUCrJ* **2022**, *9*, 682–694. [CrossRef]
40. Milano, S.K.; Huang, Q.; Nguyen, T.T.T.; Ramachandran, S.; Finke, A.; Kriksunov, I.; Schuller, D.J.; Szebenyi, D.M.; Arenholz, E.; McDermott, L.A.; et al. New Insights into the Molecular Mechanisms of Glutaminase C Inhibitors in Cancer Cells Using Serial Room Temperature Crystallography. *J. Biol. Chem.* **2022**, *298*, 101535. [CrossRef] [PubMed]
41. Stachowski, T.R.; Vanarotti, M.; Seetharaman, J.; Lopez, K.; Fischer, M. Water Networks Repopulate Protein–Ligand Interfaces with Temperature. *Angew. Chem.* **2022**, *134*, e202112919. [CrossRef]
42. Yabukarski, F.; Doukov, T.; Mokhtari, D.A.; Du, S.; Herschlag, D. Evaluating the Impact of X-Ray Damage on Conformational Heterogeneity in Room-Temperature (277 K) and Cryo-Cooled Protein Crystals. *Acta Crystallogr. Sect. D Struct. Biol.* **2022**, *78*, 945–963. [CrossRef]
43. Ayan, E.; Yuksel, B.; Destan, E.; Ertem, F.B.; Yildirim, G.; Eren, M.; Yefanov, O.M.; Barty, A.; Tolstikova, A.; Ketawala, G.K.; et al. Cooperative Allostery and Structural Dynamics of Streptavidin at Cryogenic- and Ambient-Temperature. *Commun. Biol.* **2022**, *5*, 73. [CrossRef]
44. Doukov, T.; Herschlag, D.; Yabukarski, F. Obtaining Anomalous and Ensemble Information from Protein Crystals from 220 K up to Physiological Temperatures. *Acta Crystallogr. Sect. D Struct. Biol.* **2023**, *79*, 212–223. [CrossRef]
45. Sharma, S.; Ebrahim, A.; Keedy, D.A. Room-Temperature Serial Synchrotron Crystallography of the Human Phosphatase PTP1B. *Acta Crystallogr. Sect. F* **2023**, *79*, 23–30. [CrossRef]
46. McLeod, M.J.; Barwell, S.A.E.; Holyoak, T.; Thorne, R.E. A Structural Perspective on the Temperature Dependent Activity of Enzymes. *Structure* **2025**, *33*, 924–934.e2. [CrossRef] [PubMed]
47. McPhillips, T.M.; McPhillips, S.E.; Chiu, H.-J.; Cohen, A.E.; Deacon, A.M.; Ellis, P.J.; Garman, E.; Gonzalez, A.; Sauter, N.K.; Phizackerley, R.P.; et al. Blu-Ice and the Distributed Control System: Software for Data Acquisition and Instrument Control at Macromolecular Crystallography Beamlines. *J. Synchrotron Radiat.* **2002**, *9*, 401–406. [CrossRef] [PubMed]
48. Winter, G.; Waterman, D.G.; Parkhurst, J.M.; Brewster, A.S.; Gildea, R.J.; Gerstel, M.; Fuentes-Montero, L.; Vollmar, M.; Michels-Clark, T.; Young, I.D.; et al. DIALS: Implementation and Evaluation of a New Integration Package. *Acta Crystallogr. Sect. D Struct. Biol.* **2018**, *74*, 85–97. [CrossRef]
49. Evans, P.R.; Murshudov, G.N. How Good Are My Data and What Is the Resolution? *Acta Crystallogr. Sect. D Biol. Crystallogr.* **2013**, *69*, 1204–1214. [CrossRef] [PubMed]

50. McCoy, A.J.; Grosse-Kunstleve, R.W.; Adams, P.D.; Winn, M.D.; Storoni, L.C.; Read, R.J. Phaser Crystallographic Software. *J. Appl. Crystallogr.* **2007**, *40*, 658–674. [CrossRef]
51. Afonine, P.V.; Grosse-Kunstleve, R.W.; Echols, N.; Headd, J.J.; Moriarty, N.W.; Mustyakimov, M.; Terwilliger, T.C.; Urzhumtsev, A.; Zwart, P.H.; Adams, P.D. Towards Automated Crystallographic Structure Refinement with Phenix.Refine. *Acta Crystallogr. Sect. D Biol. Crystallogr.* **2012**, *68*, 352–367. [CrossRef]
52. Emsley, P.; Lohkamp, B.; Scott, W.G.; Cowtan, K. Features and Development of Coot. *Acta Crystallogr. Sect. D Biol. Crystallogr.* **2010**, *66*, 486–501. [CrossRef]
53. Lang, P.T.; Ng, H.-L.; Fraser, J.S.; Corn, J.E.; Echols, N.; Sales, M.; Holton, J.M.; Alber, T. Automated Electron-Density Sampling Reveals Widespread Conformational Polymorphism in Proteins. *Protein Sci.* **2010**, *19*, 1420–1431. [CrossRef]
54. Ginn, H.M. Torsion Angles to Map and Visualize the Conformational Space of a Protein. *Protein Sci.* **2023**, *32*, e4608. [CrossRef]
55. Tian, H.; Xiao, S.; Jiang, X.; Tao, P. PASSer: Fast and Accurate Prediction of Protein Allosteric Sites. *Nucleic Acids Res.* **2023**, *51*, W427–W431. [CrossRef] [PubMed]
56. Ploscariu, N.; Burnley, T.; Gros, P.; Pearce, N.M. Improving Sampling of Crystallographic Disorder in Ensemble Refinement. *Acta Crystallogr. Sect. D Struct. Biol.* **2021**, *77*, 1357–1364. [CrossRef] [PubMed]
57. Schrödinger, LLC. *The PyMOL Molecular Graphics System, Version 2.5.5*; Schrödinger, LLC: New York, NY, USA, 2015.
58. Morin, A.; Eisenbraun, B.; Key, J.; Sanschagrin, P.C.; Timony, M.A.; Ottaviano, M.; Sliz, P. Collaboration Gets the Most out of Software. *eLife* **2013**, *2*, e01456. [CrossRef] [PubMed]
59. Russi, S.; Juers, D.H.; Sanchez-Weatherby, J.; Pellegrini, E.; Mossou, E.; Forsyth, V.T.; Huet, J.; Gobbo, A.; Felisaz, F.; Moya, R.; et al. Inducing Phase Changes in Crystals of Macromolecules: Status and Perspectives for Controlled Crystal Dehydration. *J. Struct. Biol.* **2011**, *175*, 236–243. [CrossRef]
60. Wheeler, M.J.; Russi, S.; Bowler, M.G.; Bowler, M.W. Measurement of the Equilibrium Relative Humidity for Common Precipitant Concentrations: Facilitating Controlled Dehydration Experiments. *Acta Crystallogr. Sect. F Struct. Biol. Cryst. Commun.* **2012**, *68*, 111–114. [CrossRef]
61. Englich, U.; Kriksunov, I.A.; Cerione, R.A.; Cook, M.J.; Gillilan, R.; Gruner, S.M.; Huang, Q.; Kim, C.U.; Miller, W.; Nielsen, S.; et al. Microcrystallography, High-Pressure Cryocooling and BioSAXS at MacCHESS. *J. Synchrotron Radiat.* **2011**, *18*, 70–73. [CrossRef]
62. Huang, Q.; Gruner, S.M.; Kim, C.U.; Mao, Y.; Wu, X.; Szebenyi, D.M.E. Reduction of Lattice Disorder in Protein Crystals by High-Pressure Cryocooling. *J. Appl. Crystallogr.* **2016**, *49*, 149–157. [CrossRef]
63. Teng, T.Y.; Moffat, K. Cooling Rates During Flash Cooling. *J. Appl. Crystallogr.* **1998**, *31*, 252–257. [CrossRef]
64. Kriminski, S.; Kazmierczak, M.; Thorne, R.E. Heat Transfer from Protein Crystals: Implications for Flash-Cooling and X-Ray Beam Heating. *Acta Crystallogr. Sect. D Biol. Crystallogr.* **2003**, *59*, 697–708. [CrossRef] [PubMed]
65. Warkentin, M.; Berejnov, V.; Husseini, N.S.; Thorne, R.E. Hyperquenching for Protein Cryocrystallography. *J. Appl. Crystallogr.* **2006**, *39*, 805–811. [CrossRef] [PubMed]
66. Clinger, J.A.; Moreau, D.W.; McLeod, M.J.; Holyoak, T.; Thorne, R.E. Millisecond Mix-and-Quench Crystallography (MMQX) Enables Time-Resolved Studies of PEPCK with Remote Data Collection. *IUCrJ* **2021**, *8*, 784–792. [CrossRef]

biomolecules

MDPI

Article

MTPrompt-PTM: A Multi-Task Method for Post-Translational Modification Prediction Using Prompt Tuning on a Structure-Aware Protein Language Model

Ye Han [1], Fei He [1], Qing Shao [2], Duolin Wang [1,*] and Dong Xu [1,*]

[1] Department of Electrical Engineering and Computer Science, Christopher S. Bond Life Sciences Center, University of Missouri, Columbia, MO 65211, USA; yhhdb@missouri.edu (Y.H.); hefe@missouri.edu (F.H.)
[2] Chemical & Materials Engineering, University of Kentucky, Lexington, KY 40506, USA; qshao@uky.edu
* Correspondence: wangdu@missouri.edu (D.W.); xudong@missouri.edu (D.X.)

Abstract: Post-translational modifications (PTMs) regulate protein function, stability, and interactions, playing essential roles in cellular signaling, localization, and disease mechanisms. Computational approaches enable scalable PTM site prediction; however, traditional models focus only on local sequence features from fragments around potential modification sites, limiting the scope of their predictions. Recently, pre-trained protein language models (PLMs) have improved PTM prediction by leveraging biological knowledge derived from extensive protein databases. However, most PLMs used for PTM site prediction are pre-trained solely on amino acid sequences, limiting their ability to capture the structural context necessary for accurate PTM site prediction. Moreover, these methods typically train separate single-task models for each PTM type, which hinders the sharing of common features and limits potential knowledge transfer across tasks. To overcome these limitations, we introduce MTPrompt-PTM, a multi-task PTM prediction framework developed by applying prompt tuning to a structure-aware protein language model (S-PLM). Instead of training several single-task models, MTPrompt-PTM trains one multi-task model to predict multiple types of PTM sites using shared feature extraction layers and task-specific classification heads. Additionally, we incorporate a knowledge distillation strategy to enhance the efficiency and generalizability of multi-task training. Experimental results demonstrate that MTPrompt-PTM outperforms state-of-the-art PTM prediction tools on 13 types of PTM sites, highlighting the advantages of multi-task learning and structural integration.

Keywords: post-translational modification prediction; multi-task prediction; prompt tuning; structure-aware protein language model (S-PLM); knowledge distillation

1. Introduction

Post-translational modifications (PTMs) are crucial regulators of protein function, stability, and interactions. These modifications occur after translation and play essential roles in cellular signaling, protein localization, and disease mechanisms [1–3]. Although over 400 distinct PTM types have been identified, most remain poorly characterized regarding their target sites and biological context [4]. Experimental techniques such as mass spectrometry (MS), Western blotting, and radiolabeling are widely used for PTM identification; however, they are expensive, time-consuming, and constrained by technical limitations [5–7]. Computational approaches address these challenges by providing fast, cost-effective, and scalable PTM site prediction [8–10]. A common practice involves training models on existing PTM datasets to identify potential PTM sites on unseen data. This

 https://doi.org/10.3390/biom15060843

approach can be broadly categorized into supervised training from scratch or fine-tuning pre-trained protein language models (PLMs).

Training models from scratch typically involves using protein sequence fragments or local structural information through machine learning or deep learning methods. For example, NetPhos 3.1 [10] developed an artificial neural network (ANN) model incorporating sequence-based motifs and structural features to predict phosphorylation sites in eukaryotic proteins. NetNGlyc 1.0 [11] built prediction models for N-linked, O-linked, and C-linked glycosylation sites by utilizing artificial neural networks that examined the sequence context and surface accessibility of potential glycosylation sites. The group-based prediction system (GPS) algorithm in GSP-MSP [12] integrates sequence features to identify specific methylation types on lysine and arginine residues in proteins. MethylSight [13] created a machine learning model that predicts lysine methylation sites in human proteins by utilizing alignment-free features that capture structural information around lysine residues. Ertelt et al. [14] combined machine learning and structure-based protein design to predict and engineer protein post-translational modifications (PTMs), offering a powerful tool for synthetic biology and therapeutic development.

In addition to traditional machine learning techniques, deep learning-based methods have been applied for PTM prediction. For example, MusiteDeep [15] uses convolutional neural networks (CNNs) to automatically learn sequence representations, overcoming the limitations of traditional feature engineering and achieving improved accuracy in phosphorylation site identification. Meanwhile, CapsNet-PTM [16] employs capsule networks to predict seven different PTM types by capturing spatial dependencies between PTM features. Additionally, Wang et al. [17] introduced a web server for the prediction and visualization of 13 PTM types by combining MusiteDeep/CNN and CapsNet deep learning networks and leveraging advanced ensemble techniques. Furthermore, GPS-SUMO 2.0 [18] utilized three advanced machine learning methods—penalized logistic regression (PLR), deep neural networks (DNNs), and Transformer models—to improve the prediction of SUMOylation sites by incorporating multiple sequence features. However, these methods focus only on local sequence features from fragments around potential modification sites, limiting the scope of their predictions.

Training from pre-trained protein language models (PLMs) has proven highly successful in predicting PTM sites. Recently, embeddings from various pre-trained PLMs, such as ESM2 [19], ProtBERT [20], and ProtT5 [20], have been used as features for the training of PTM site prediction models. For instance, Lmnglypred [21] utilizes ProtT5 embeddings to predict N-linked glycosylation sites. PTG-PLM [22] uses embeddings from multiple PLMs, including ProtBERT-BFD, ProtAlbert, ProtXLNet, ESM-1b, and TAPE, to enhance glycosylation and glycation site prediction using CNNs. LM-OGlcNAc-Site [23] applies sophisticated ensemble strategies by combining embeddings from Ankh, ESM-2, and ProtT5 to predict O-linked N-acetylglucosamine (O-GlcNAc) modification sites. In contrast to the aforementioned methods, which rely solely on PLM embeddings as features, PTM-GPT2 [24] fine-tunes a decoder-based autoregressive Transformer model, ProtGPT2, using a custom prompt to guide the model in accurately predicting PTM sites. More recently, PTM-Mamba [25] introduces a novel protein language model that integrates PTM information by incorporating PTM-specific tokens through bidirectional Mamba blocks and fusing them with ESM-2 embeddings using a gating mechanism. The resulting representations can be directly applied to downstream tasks such as phosphorylation and non-histone acetylation site prediction. In contrast to training models from scratch, these methods benefit from embeddings pre-trained on extensive protein sequence databases, allowing them to capture both local sequence motifs (e.g., short patterns around modification sites) and the global sequence context (e.g., long-range dependencies in protein sequences). This makes them highly effective for PTM site prediction. However, these methods have several limitations. First, 3D structural information plays a crucial role in PTM prediction, as most PTMs occur at solvent-exposed residues rather than buried ones, and PTM sites are often influenced by non-local sequence interactions due to protein folding [26]. However, most PLMs used for PTM site prediction are trained solely on amino acid sequences, limiting their ability to capture the structural context necessary for accurate PTM site prediction. Second, different PTM types often occur close to each other on the protein sequence and can share sequence motifs and structural dependencies. However, these methods typically train models for different PTM types separately, preventing the sharing of common features among them. The advantages and disadvantages of the representative PTM prediction tools are presented in Table 1.

Table 1. Overview of representative PTM prediction tools.

Category	Model	Description	Advantages	Disadvantages
Machine Learning-Based	NetPhos 3.1 NetNGlyc 1.0 GPS-MSP MethylSight Ertelt et al.	Use manually designed features with classical ML models such as ANNs or SVM.	Easy to interpret and efficient for small datasets, producing well-established tools in early PTM prediction research.	Cannot capture long-range dependencies, rely heavily on expert-crafted features, and generalize poorly to unseen data.
Deep Learning-Based	MusiteDeep CapsNet-PTM GPS-SUMO 2.0	Leverage CNNs, CapsuleNets, and other DL architectures to automatically learn features from sequence data.	Automatically learn features from raw data and offer better performance on large-scale datasets.	Rely on local sequence windows, ignore structural information, are usually trained separately for each PTM type, and require large, labeled datasets.
Protein Language Model-Based	Lmnglypred PTG-PLM O-GlcNAc PTM-GPT2 PTM-Mamba	Use embeddings from large-scale pre-trained PLMs or fine-tune PLMs for PTM prediction.	Capture long-range sequence dependencies, benefit from massive pre-training, support transfer learning and generalization.	Lack of direct structural context and rarely leverage effective joint learning across multiple PTM types.

To address the limitations mentioned above, we propose MTPrompt-PTM, a novel multi-task PTM site prediction model that leverages the prompt tuning of a structure-aware protein language model (S-PLM) [27] for 13 types of PTM sites, including phosphorylation (S, T, Y), N-linked glycosylation (N), O-linked glycosylation (S, T), ubiquitination (K), acetylation (K), methylation (K, R), SUMOylation (K), succinylation (K), and palmitoylation (C). Our model consists of an encoder and a decoder. The encoder uses S-PLM as the backbone to encode the protein sequence. S-PLM is a pre-trained PLM incorporating structural information with sequence-based embeddings from ESM2. During the multi-task training phase, all parameters of S-PLM are frozen. However, to effectively leverage the pre-trained model's information, we perform prompt tuning on S-PLM. Prompt tuning [28] is a parameter-efficient fine-tuning (PEFT) technique that adds trainable embeddings, called 'prompts', to the sequence embeddings. Unlike full fine-tuning, where all model weights are updated, prompt tuning optimizes only the additional trainable embeddings, reducing the computational overhead while maintaining generalization. In our model, we propose a novel method for the initialization of our task prompts. The decoder consists of shared layers and task-specific layers, which capture common features and task-specific features separately. Additionally, we incorporate a knowledge distillation strategy, where single-task models teach a multi-task model, helping the multi-task model to outperform its single-task counterparts by integrating knowledge across multiple tasks. Our experimental results show that MTPrompt-PTM improves the predictive performance compared to single-task models. To further validate its effectiveness, we compare MTPrompt-PTM with state-of-the-art PTM prediction tools. The results demonstrate that MTPrompt-PTM outperforms these tools across all 13 PTM types, confirming its effectiveness.

2. Materials and Methods

2.1. Dataset and Data Processing

Numerous databases and research studies provide PTM data; however, most of them offer only short peptide fragments centered around the modified residue, lacking a full protein context or complete sequence information. These truncated sequences often miss the essential global context, making it difficult to capture long-range interactions between residues, which can be critical in determining PTM occurrence. To address this limitation, we utilize full-length protein sequences in our study, enabling the model to capture the comprehensive contextual information and long-range sequence dependencies necessary for accurate PTM site prediction.

UniProt, the largest protein sequence database with PTM annotations, contains over 200 million protein sequences and provides annotations for 200 PTM types. Therefore, we constructed a new PTM dataset from UniProt, incorporating 13 PTM types: phosphorylation (S, T, Y), N-linked glycosylation (N), O-linked glycosylation (S, T), ubiquitination (K), acetylation (K), methylation (K, R), SUMOylation (K), succinylation (K), and palmitoylation (C).

To build this dataset, we first downloaded full-length protein sequences along with their PTM annotations for the 13 PTM types from UniProt. The data were then filtered by species and sequence length, retaining only metazoan proteins and excluding sequences longer than 1022 residues. Sequences longer than 1022 were not processed because longer sequences would have exceeded the model's input size limit and negatively affected the Transformer model's performance. Processing excessively long sequences can lead to memory overload, slower processing times, and reduced effectiveness due to the quadratic complexity ($O(n^2)$) of the attention mechanism. We chose not to truncate the sequences because truncation could result in losing important functional or structural information, especially for long protein sequences, where key motifs or functional sites might be outside the truncated region. By limiting the sequence length to 1022 tokens, we ensured that we retained the most relevant parts of the sequence without losing critical details, striking a balance between computational efficiency and maintaining the essential sequence context. Table 2 presents the PTM types, the corresponding UniProt annotations, and the number of protein sequences.

Table 2. UniProt annotations and number of downloaded protein sequences for different PTM types.

PTM Type	PTM Annotation in UniProt	Number of Protein Sequences
Phosphorylation (S)	Phosphoserine; Diphosphoserine; O-(2-cholinephosphoryl)serine; (Microbial infection) Phosphoserine; O-(pantetheine4′phosphoryl)serine; (Microbial infection) O-(2-cholinephosphoryl) serine	12,230
Phosphorylation (T)	(Microbial infection) Phosphothreonine; Phosphothreonine	8551
Phosphorylation (Y)	Phosphotyrosine	3782
Ubiquitination (K)	(Microbial infection) Glycyl lysine isopeptide (Lys-Gly) (interchain with G-Cter in ubiquitin); Glycyl lysine isopeptide (Lys-Gly) (interchain with G-Cter in ubiquitin and interchain with MARCHF2); Glycyl lysine isopeptide (Lys-Gly) (interchain with G-Cter in ubiquitin)	1225

Table 2. *Cont.*

PTM Type	PTM Annotation in UniProt	Number of Protein Sequences
N-Linked Glycosylation (N)	N-linked (GlcNAc...) (paucimannose) asparagine; N-linked (GlcNAc...) (keratan sulfate) asparagine; N-linked (GlcNAc...) (complex) asparagine; N-linked (GlcNAc) asparagine; N-linked (Glc...) asparagine; N-linked (GlcNAc...) (hybrid) asparagine; N-linked (GalNAc...) asparagine; N-linked (GlcNAc...) (polylactosaminoglycan) asparagine; N-linked (GlcNAc...) asparagine; N-linked (Hex) asparagine; N-linked (HexNAc...) asparagine; N-linked (GlcNAc...) (high mannose) asparagine	12,285
O-Linked Glycosylation (S)	O-linked (Xyl...) (dermatan sulfate) serine; O-linked (Fuc...) serine; O-linked (Xyl...) (heparan sulfate) serine; O-linked (HexNAc...) serine; O-linked (Fuc) serine; O-linked (GalNAc...) serine; O-linked (Xyl...) serine; O-linked (Hex...) serine; O-linked (GlcA) serine; O-linked (GlcNAc) serine; O-linked (GalNAc) serine; O-linked (Man...) serine; O-linked (Xyl...) (glycosaminoglycan) serine; O-linked (Hex) serine; O-linked (GlcNAc...) serine; O-linked (Glc...) serine; O-linked (Xyl...) (chondroitin sulfate) serine; O-linked (Man) serine	942
O-Linked Glycosylation (T)	O-linked (GlcNAc...) threonine; O-linked (Xyl...) (keratan sulfate) threonine; O-linked (Hex) threonine; O-linked (GalNAc) threonine; O-linked (GalNAc...) threonine; O-linked (GlcNAc) threonine; (Microbial infection) O-linked (Glc) threonine; O-linked (Fuc) threonine; O-linked (HexNAc) threonine; O-linked (Man6P...) threonine; O-linked (Man...) threonine; O-linked (Fuc...) threonine; O-linked (HexNAc...) threonine; O-linked (Hex...) threonine; O-linked (Man) threonine	694
Acetylation (K)	N6-acetyllysine; N6-acetyl-N6-methyllysine; (Microbial infection) N6-acetyllysine	6009
Palmitoylation (C)	N-palmitoyl cysteine; S-palmitoyl cysteine	1531
Methylation (R)	Asymmetric dimethylarginine; N5-[4-(S-L-cysteinyl)-5-methyl-1H-imidazol-2-yl]-L-ornithine (Arg-Cys) (interchain with C-151 in KEAP1); Symmetric dimethylarginine; Dimethylated arginine; Omega-N-methylated arginine; Omega-N-methylarginine	1680
Methylation (K)	N6-acetyl-N6-methyllysine; N6-methyllysine; N6,N6,N6-trimethyllysine; N6-methylated lysine; N6,N6-dimethyllysine	578
SUMOylation (K)	Glycyl lysine isopeptide (Lys-Gly) (interchain with G-Cter in SUMO; Glycyl lysine isopeptide (Lys-Gly) (interchain with G-Cter in SUMO1, SUMO2 and SUMO3); Glycyl lysine isopeptide (Lys-Gly) (interchain with G-Cter in /SUMO5); Glycyl lysine isopeptide (Lys-Gly) (interchain with G-Cter in SUMO); Glycyl lysine isopeptide (Lys-Gly) (interchain with G-Cter in SUMO3); Glycyl lysine isopeptide (Lys-Gly) (interchain with G-Cter in SUMO1); Glycyl lysine isopeptide (Lys-Gly) (interchain with G-Cter in SUMO2 and SUMO3); Glycyl lysine isopeptide (Lys-Gly) (interchain with G-Cter in SUMO1 and SUMO2); Glycyl lysine isopeptide (Lys-Gly) (interchain with G-Cter in SUMO1P1/SUMO5); Glycyl lysine isopeptide (Lys-Gly) (interchain with G-Cter in SUMO2)	3724
Succinylation (K)	N6-succinyllysine	2069

For each protein sequence, the PTM sites are treated as positive samples, while other positions with the same amino acids, excluding the PTM sites, are treated as negative samples. Figure 1 shows the number of PTM sites in terms of positive and negative sites for each PTM type. During training, the entire protein sequence is input, but the loss is calculated only for the positive and negative sites.

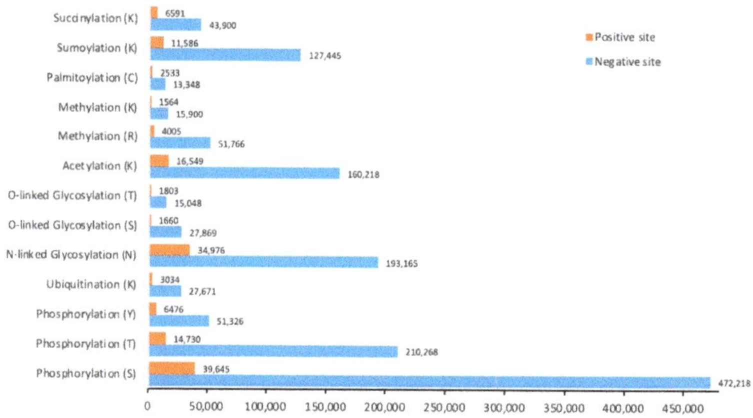

Figure 1. The distribution of positive and negative PTM sites across different PTM types.

We then separated all the protein sequences into a training set and a testing set based on the timestamp. Protein sequences annotated in UniProt prior to 2010 were used for training, while those annotated after 2010 were reserved for testing. We trained our model on the training data and used the test data to compare our model's performance with that of other state-of-the-art tools. Additionally, we applied the widely used clustering program CD-HIT-2D to assess the similarity between the training and testing data. The testing protein sequences with no more than 60%, 70%, and 80% similarity to the training data were generated using CD-HIT-2D. We present the performance of the testing data at different levels of sequence similarity to the training data.

Furthermore, we created another non-redundant dataset to evaluate our model. We applied CD-HIT [29] to cluster this dataset based on a 60% sequence similarity threshold. To avoid homologous redundancy, only one representative sequence from each cluster was selected. The non-redundant dataset was then split into training and testing sets in a 4:1 ratio. These datasets were used to train and evaluate our model, ensuring a robust performance assessment.

We further evaluated our model using an independent benchmark for phosphorylation and non-histone acetylation. The phosphorylation test set was obtained from the ProteinBERT benchmark [30], which is derived from PhosphoSitePlus [31], a comprehensive resource of experimentally validated post-translational modifications in human and mouse proteins. The non-histone acetylation test set was sourced from TransPTM [32], a Transformer-based model specifically designed for the prediction of non-histone acetylation sites.

2.2. *Architecture of MTPrompt-PTM*

This paper introduces MTPrompt-PTM, a multi-task model for post-translational modification prediction. The overall architecture of MTPrompt-PTM is illustrated in Figure 2. Our model includes an encoder and decoder. The encoder leverages S-PLM v2 as its backbone and is trained using prompt tuning with task prompts. The decoder is a hybrid architecture comprising shared feature extraction layers and task-specific classification layers.

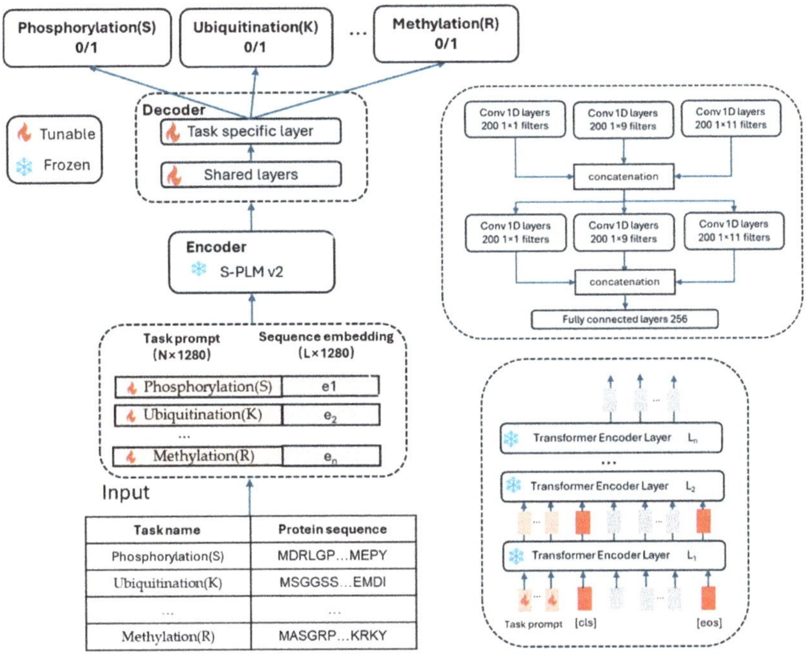

Figure 2. The architecture of MTPrompt-PTM. During multi-task training, the model takes the task name and protein sequence as input. The task prompts are initialized using our proposed method based on the task name. The protein sequence is tokenized and embedded using the ESM2 tokenizer. The task prompts are then concatenated with the sequence embeddings and input into the encoder. The backbone of the encoder is S-PLM v2, with the input passing through 33 Transformer encoder layers. Then, residue-level representations (excluding [CLS], [EOS], and task tokens) are extracted and passed to the decoder. During the entire training process, while the parameters of S-PLM remain frozen, the task prompts are updated through gradient descent. The decoder features a hybrid architecture with shared and task-specific layers. The shared component consists of two CNN Inception modules, each containing three 1D convolutional layers with varying kernel sizes, followed by concatenation and a fully connected layer. The task-specific layers process the shared residue representation and perform classification. Each task-specific head corresponds to a different PTM type, receiving residue representations and outputting whether the residue belongs to the respective PTM type.

Unlike most PTM prediction methods that use peptides as input, our model takes entire protein sequences as input. Initially, the protein sequences containing PTM sites are tokenized using the ESM2 tokenizer, which converts them into sequence embeddings. Task prompts, which act as additional trainable embeddings to guide the model in distinguishing between different PTM prediction tasks, are concatenated with the protein sequence embedding. This combined matrix is then passed through the encoder as usual. Throughout the multiple Transformer layers in S-PLM v2, the model generates updated task prompts and protein sequence embeddings. After the Transformer layers, the task prompts, along with the [CLS] and [EOS] tokens, are discarded. [CLS] (Classification) and [EOS] (End of Sequence) are special tokens commonly used in Transformer-based language models. The [CLS] token is typically added at the beginning of an input sequence, and its corresponding output embedding is used for classification tasks, summarizing the entire input. The [EOS] token marks the end of a sequence, signaling the model where the input terminates, which is particularly important in generative or sequential prediction tasks.

Only the residue-level embeddings for the sequence are retained and passed to the decoder for further processing. During the entire training process, while the parameters of S-PLM remain frozen, the task embeddings are updated through gradient descent.

Our decoder is designed with both shared layers and task-specific layers. The shared layers consist of two CNN Inception modules and a fully connected (FC) layer. Each CNN Inception module is composed of three 1D CNN layers with different filter sizes, enabling the capture of multi-dimensional local information from the input sequences. The outputs from these two modules are concatenated to combine the captured local features. Following the CNN layers, a fully connected layer is added to further capture and refine the information. These shared layers are responsible for learning common, generalizable representations that can be applied across different post-translational modification (PTM) types. Once the shared representation is learned, task-specific classification layers are introduced to handle the unique characteristics of each PTM type. These task-specific layers consist of 13 fully connected layers, each corresponding to a different PTM type. These layers can be seen as 13 independent classification heads, with each head trained to focus on the specific sequence patterns, structural features, or biochemical properties associated with its respective PTM. Each PTM-specific head independently predicts the probability of the presence or absence of its respective modification at each relevant sequence position. By maintaining separate classification heads for each PTM type, the model ensures that features are tailored for each modification, enhancing the model's predictive accuracy and allowing for more precise modeling of the diverse PTM signals.

2.3. *Prompt Tuning on MTPrompt-PTM*

In our encoder, to generate residue-level embeddings with enhanced structural information, we use S-PLM v2 [33] as the backbone. S-PLM [27] is a structure-aware protein language model integrating both sequence and structural information via contrastive learning. S-PLM v2 is the upgraded version of S-PLM, using a geometric vector perceptron (GVP) model [34] to achieve more precise residue-level embeddings by capturing detailed geometric properties. The sequence encoder of S-PLM builds upon a pre-trained ESM2 model, preserving previously learned protein knowledge while effectively adapting to new tasks. In contrast to ESM2, S-PLM explicitly incorporates structural information due to its pre-training on paired protein sequences and contact maps, enabling the direct encoding of spatial relationships and residue–residue interactions into its representations. T-SNE clustering results have shown that S-PLM achieves superior kinase group clustering compared to ESM2, underscoring its potential as an effective backbone model for PTM site prediction [27].

Although S-PLM already contains rich general knowledge learned from large-scale protein sequence data, we use prompt tuning to make task-specific adjustments for the prediction of post-translational modifications. The core idea of prompt tuning is to concatenate the task prompts with the protein sequence embeddings and input them into the pre-trained language model. This allows the Transformer operations to be performed while keeping the original model's weights frozen. As a result, the final protein sequence embeddings are adjusted by the task prompts. Given that prompt tuning is highly sensitive to the initialization of the task prompts, it is crucial to initialize the prompts effectively. Therefore, we propose a novel initialization method for different tasks. First, we collect all the protein sequences from the training set and obtain their sequence embeddings by inputting them into S-PLM v2. Next, we extract 21-residue peptides centered on the PTM site and compute the average of their embeddings. These averages are then clustered into K clusters. Finally, we average the values within each cluster to generate the final prompt matrix. The specific initialization process is outlined below.

Step 1. Extracting Protein Embeddings

We utilized S-PLM v2 to generate embeddings for every residue in the training protein sequences. By feeding the entire training set into S-PLM, we obtained residue-level embeddings with a shape of $N \times 1280$, where N represents the total number of residues. The value 1280 corresponds to the dimensionality of the embedding vector produced by S-PLM v2 for each residue. These embeddings capture rich, context-aware biochemical and structural information for each amino acid, providing a robust foundation for downstream PTM prediction.

Step 2. Generating PTM-Centered Embeddings

Since PTMs are often influenced by the local sequence environment surrounding the modified residue, we extract a 21-residue window centered on each PTM site to capture this context. This window includes a modified residue along with its ten upstream and ten downstream neighbors, effectively preserving the immediate biochemical environment. For each PTM site, the contextual window S_i is defined as

$$S_i = \{E_{i-10}, E_{i-9}, \ldots, E_i, \ldots, E_{i+9}E_{i-10}\},\ S_i \in R^{21\times 1280} \quad (1)$$

where E_i represents the embedding of the i-th residue. These windows provide a localized, high-dimensional representation of the sequence, which is essential for accurate PTM site modeling.

Step 3. Computing Mean Embeddings

Each window is then averaged to produce a single 1280-dimensional vector that represents the local environment of the PTM site. The mean embedding E_i for each PTM site is computed as

$$E_i = \frac{1}{21}\sum\nolimits_{j=i-10}^{i+10} E_j,\ E_i \in R^{1280} \quad (2)$$

This mean pooling simplifies the representation while retaining the essential pattern of this PTM context.

Step 4. Clustering PTM Representations

Instead of averaging all PTM site embeddings into a single global representation, we apply K-means clustering to group them into K distinct clusters. After experimenting with different values of K, we found that the best results were achieved when $K = 500$. Each cluster captures a recurring motif or feature pattern shared across different proteins, preserving the diversity and subtlety of PTM-specific contexts. Formally, each cluster C_k is defined as

$$C_k = \{E_i | i \in cluster\ k\} \quad (3)$$

The rationale behind selecting K clusters is to preserve the inherent diversity and fine-grained patterns captured in the embedding space. Biological modifications (PTMs) often exhibit subtle yet meaningful variations, reflecting different functional contexts, regulatory mechanisms, or kinase substrate specificity. By clustering the embeddings into multiple distinct groups, we maintain these biologically relevant variations, allowing each cluster to represent a unique pattern or functional state more accurately. Direct averaging could obscure these subtle differences and lead to the loss of critical biological insights.

Step 5. Computing Cluster Centroids

For the k-th cluster, the centroid vector (the mean of the embeddings in this cluster) μ_k is computed as

$$\mu_k = \frac{1}{|C_k|}\sum\nolimits_{E_i \in C_k} E_i,\quad \mu^k \epsilon R^{1280} \quad (4)$$

These centroids serve as prototypical representations of common PTM-related contexts.

Step 6. Constructing the Final Prompt Matrix

The 500 centroids obtained from clustering are then stacked to form a matrix that serves as a task-specific embedding. This matrix acts as a learnable guide for the model, capturing diverse PTM-related patterns and functional contexts. The final prompt embedding matrix M has a shape of $K \times 1280$ and is defined as

$$M = [\mu_1; \mu_2; \ldots; \mu_K] \epsilon R^{K \times 1280} \quad (5)$$

2.4. Multi-Task Training of MTPrompt-PTM

To improve both the performance and generalizability of the multi-task model, we adopt a knowledge distillation strategy, which transfers knowledge from a teacher model to a student model by training the student to imitate the teacher's outputs. Unlike traditional training with one-hot labels, the teacher's probability distribution over classes, referred to as soft labels, provides a richer and more informative learning signal. Clark et al. [35] demonstrated that using a single-task teacher model to guide a multi-task student model is significantly more effective than employing multiple teachers for multiple tasks. This is because the student benefits from exposure to a diverse set of PTM-specific teachers, similarly to how ensemble learning enhances generalization. Inspired by this, we apply knowledge distillation in our framework by using single-task models as teachers to train the multi-task model, enabling it to leverage both expert knowledge and shared task representations for improved performance.

As shown in Figure 3, the training process consists of two main steps. In Step 1, we independently train 13 single-task models for all 13 PTM types. These single-task models act as teacher models, with architectures nearly identical to that of the multi-task model, except for the absence of task-specific layers. After training, we use the single-task models to generate predictions for all training data. These predictions, serving as soft labels, capture subtle patterns and uncertainties often missed by traditional hard labels. In Step 2, we merge the training data from all PTM types to train the multi-task student model. The soft labels from the teacher models are used to guide the student model's training. To further enhance the training, we adopt a teacher annealing strategy [35], progressively blending the teacher's soft labels with ground truth annotations. This combination provides a refined supervisory signal, improving the model's generalization and accuracy across diverse PTM types. To address potential class imbalances from simply concatenating all datasets, we apply a weighted loss function that combines soft labels (from teacher models) and hard labels (ground truth annotations). The weights are determined based on the dataset size, and the loss function is defined as

$$\mathrm{L}(\theta) = \sum_{t \epsilon T} weight_t \sum_{x_t^i y_t^i \epsilon D_t} l\left(\gamma y_t^i + (1-\gamma) f_t\left(x_t^i, \theta_t\right), f_t\left(x_t^i, \theta\right)\right), \ \gamma \in (0,1) \quad (6)$$

Here, T denotes the set of PTM tasks. For each task t, we first train a single-task teacher model with parameters θ_t and then use its predictions to guide the multi-task student model with parameters θ. The variable γ controls the balance between hard and soft labels during training; specifically, we set $\gamma = 0.5$ to dynamically reconstruct the training labels by averaging the teacher's soft predictions and the ground truth.

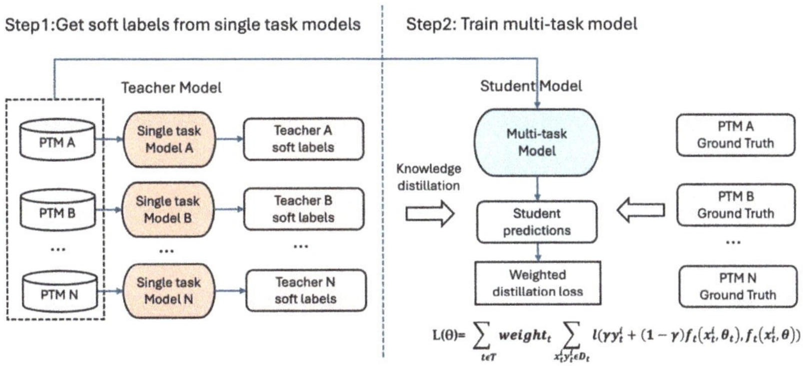

Figure 3. Training process of MTPrompt-PTM. In Step 1, we independently train 13 single-task models, each corresponding to a different PTM type, which serve as teacher models. The saturated orange ovals are teacher models. These models have architectures similar to that of the multi-task model, with the key difference being the absence of task-specific layers. The teacher models generate soft labels through predictions on the training data, capturing subtle patterns and uncertainties missed by traditional hard labels. The paler peach boxes immediately to their right are the soft-label outputs those teachers produce. In Step 2, we merge the training data from all PTM types to train the multi-task student model. The powder-blue oval is the student model that learns all tasks jointly. The soft labels from the teacher models guide the student model's training. A teacher annealing strategy is applied, progressively blending the teacher's soft labels with ground truth annotations to improve the model's generalization and accuracy. The light-blue rectangle below it is the student's prediction. To address class imbalances, we use a weighted loss function that combines both soft labels and hard labels, with weights determined by the dataset size.

3. Results

3.1. Comparison with State-of-the-Art Tools

Here, we compare MTPrompt-PTM with several state-of-the-art PTM prediction tools, including MusiteDeep [17], PTMGPT2 [24], NetPhos3.1 [10], NetOGlyc4.0 [36], NetNGlyc1.0 [11], GPS-SUMO2.0 [18], CSS-Palm4.0 [37,38], GSP-MSP [12], and MethylSight [13]. Table 2 presents the prediction results on the test set for all 13 PTM types, including phosphorylation (S, T, Y), N-linked glycosylation (N), O-linked glycosylation (S, T), ubiquitination (K), acetylation (K), methylation (K, R), SUMOylation (K), succinylation (K), and palmitoylation (C).

As shown in Table 3, MTPrompt-PTM outperformed all other tools across all 13 PTM types. For example, in terms of the Matthews correlation coefficient (MCC), phosphorylation (S) exhibited a substantial improvement, with MTPrompt-PTM achieving a 118.9% increase in the MCC compared to MusiteDeep. This trend continued with phosphorylation (T, Y), where our model outperformed MusiteDeep by 58.0% and 26.5%, respectively. In the case of O-linked glycosylation (S, T), MTPrompt-PTM showed improvements of 24.2% and 63.6% over MusiteDeep. For N-linked glycosylation (N), MTPrompt-PTM showed a smaller improvement of 3.2%. For SUMOylation (K), MTPrompt-PTM outperformed PTMGPT2 by 46.2%, while MusiteDeep did not achieve comparable results. This difference can be attributed to the fact that MusiteDeep uses a smaller dataset than the one used to train our model. Similarly, for ubiquitination (K) and acetylation (K), our model exceeded MusiteDeep by 140.3% and 16.7%, respectively. Since MusiteDeep does not provide a model for the prediction of succinylation, we compared MTPrompt-PTM only with PTMGPT2 for this PTM type. Here, our model achieved a 104.2% improvement over PTMGPT2 in succinylation (K). For palmitoylation (C), methylation (R), and methylation

(K), MTPrompt-PTM surpassed MusiteDeep by 28.5%, 50.7%, and 13.3%, respectively. These results demonstrate that our model outperforms many existing PTM prediction tools and can be considered a leading tool for PTM site prediction.

Table 3. Performance comparison between different methods.

PTM Type	Method	Accuracy	F1	MCC	Precision	Recall
Phosphorylation (S)	MTPrompt-PTM	**0.964**	**0.736**	**0.718**	**0.772**	0.704
	MusiteDeep	0.706	0.311	0.328	0.187	**0.917**
	PTMGPT2	0.821	0.281	0.225	0.198	0.483
	NetPhos3.1	0.289	0.155	0.09	0.085	0.905
Phosphorylation (T)	MTPrompt-PTM	**0.975**	**0.811**	**0.798**	**0.787**	**0.836**
	MusiteDeep	0.903	0.512	0.505	0.378	0.793
	PTMGPT2	0.88	0.32	0.272	0.252	0.439
	NetPhos3.1	0.429	0.153	0.103	0.085	0.802
Phosphorylation (Y)	MTPrompt-PTM	**0.973**	**0.873**	**0.858**	**0.877**	**0.87**
	MusiteDeep	0.921	0.705	0.678	0.593	0.87
	PTMGPT2	0.785	0.407	0.342	0.29	0.681
	NetPhos3.1	0.613	0.262	0.154	0.165	0.634
O-Linked Glycosylation (S)	MTPrompt-PTM	**0.983**	**0.774**	**0.765**	**0.762**	0.787
	MusiteDeep	0.956	0.6	0.616	0.454	**0.885**
	PTMGPT2	0.929	0.363	0.351	0.273	0.541
	NetOGlyc4.0	0.718	0.151	0.163	0.085	0.672
O-Linked Glycosylation (T)	MTPrompt-PTM	**0.962**	**0.786**	**0.769**	**0.724**	**0.859**
	MusiteDeep	0.87	0.49	0.47	0.357	0.781
	PTMGPT2	0.892	0.386	0.329	0.355	0.422
	NetOGlyc4.0	0.801	0.267	0.196	0.19	0.453
N-Linked Glycosylation (N)	MTPrompt-PTM	**0.977**	**0.915**	**0.903**	**0.889**	0.944
	MusiteDeep	0.967	0.887	0.875	0.802	**0.992**
	PTMGPT2	0.944	0.808	0.782	0.73	0.905
	NetNGlyc1.0	0.294	0.233	0.033	0.136	0.825
SUMOylation (K)	MTPrompt-PTM	**0.97**	**0.838**	**0.823**	**0.878**	0.802
	MusiteDeep	0.901	0.332	0.299	0.472	0.256
	PTMGPT2	0.92	0.606	0.563	0.572	0.644
	GPS-SUMO2.0	0.187	0.187	0.081	0.103	**0.978**
Ubiquitination (K)	MTPrompt-PTM	**0.956**	**0.782**	**0.764**	**0.879**	0.704
	MusiteDeep	0.681	0.367	0.318	0.236	**0.831**
	PTMGPT2	0.634	0.259	0.14	0.167	0.575
Succinylation (K)	MTPrompt-PTM	**0.983**	**0.933**	**0.923**	**0.944**	**0.922**
	PTMGPT2	0.834	0.519	0.452	0.408	0.715
Acetylation (K)	MTPrompt-PTM	**0.981**	**0.899**	**0.889**	**0.924**	0.877
	MusiteDeep	0.948	0.778	0.762	0.671	**0.926**
	PTMGPT2	0.724	0.286	0.199	0.192	0.562
Palmitoylation (C)	MTPrompt-PTM	**0.974**	**0.926**	**0.91**	**0.943**	0.909
	MusiteDeep	0.916	0.759	0.708	0.774	0.745
	PTMGPT2	0.883	0.695	0.626	0.651	0.745
	CSS-Palm4.0	0.183	0.3	-0.03	0.177	**0.982**

Table 3. *Cont.*

PTM Type	Method	Accuracy	F1	MCC	Precision	Recall
Methylation (R)	MTPrompt-PTM	**0.989**	**0.892**	**0.889**	**0.967**	0.829
	MusiteDeep	0.915	0.565	0.59	0.399	**0.967**
	PTMGPT2	0.921	0.54	0.54	0.403	0.819
	GSP-MSP	0.734	0.274	0.304	0.162	0.881
Methylation (K)	MTPrompt-PTM	**0.976**	**0.883**	**0.87**	**0.901**	0.867
	MusiteDeep	0.952	0.791	0.768	0.728	0.867
	PTMGPT2	0.869	0.529	0.477	0.427	0.695
	GSP-MSP	0.337	0.234	0.162	0.133	**0.962**
	MethylSight	0.384	0.19	0.022	0.11	**0.686**

Note: Numbers in bold represent the highest values achieved for each metric (Accuracy, F1, MCC, Precision, Recall) within a given PTM type.

Figures 4 and 5 present the ROC curves and precision–recall curves on the test set for all PTM types across different methods. Since PTM-GPT2 only provides binary predictions (i.e., whether a residue is a PTM site or not) without probability scores, we could not plot its full ROC and precision–recall curves. However, we calculated its TPR, FPR, precision, and recall to mark PTM-GPT2 as a single point on the curves. The AUC and PRAUC values of all methods are consistent with the results in Table 2, further demonstrating that our model outperforms other approaches.

To further evaluate the kinase-level specificity and robustness of MTPrompt-PTM, we conducted a comparative analysis across several kinase families. Table 4 summarizes the number of kinase sites included in both the training and testing sets, as well as the number of correctly predicted kinase sites and the corresponding accuracy for each model. Our results demonstrate that MTPrompt-PTM consistently achieves high accuracy across most kinase families. Particularly in the CMGC and AGC families, which have relatively large numbers of kinase sites in both the training and testing sets, MTPrompt-PTM significantly outperforms other models, indicating its strong generalization abilities in data-rich settings. These findings suggest that our multi-task prompt tuning framework and structure-aware backbone not only improve overall PTM prediction but also enhance kinase-specific site recognition. This highlights the potential of MTPrompt-PTM as a generalizable tool for kinase-centered PTM analysis. Notably, all models perform poorly on the "other" kinase family, with MTPrompt-PTM achieving accuracy of only 0.111 and the remaining models showing similarly low or inconsistent performance. This may be attributed to the relatively limited number of training samples available for this group (only 415 sites, compared to over 1700 for CMGC or 879 for AGC), which constrains the model's ability to learn meaningful and generalizable features for these kinases. Additionally, the "other" category likely includes a diverse and heterogeneous set of kinases that do not share common sequence or structural motifs, further complicating the prediction task. This observation highlights a common challenge in PTM prediction: models tend to favor well-represented kinase families during training (e.g., AGC and CMGC), and their performance may degrade when applied to underrepresented or diverse groups (e.g., other).

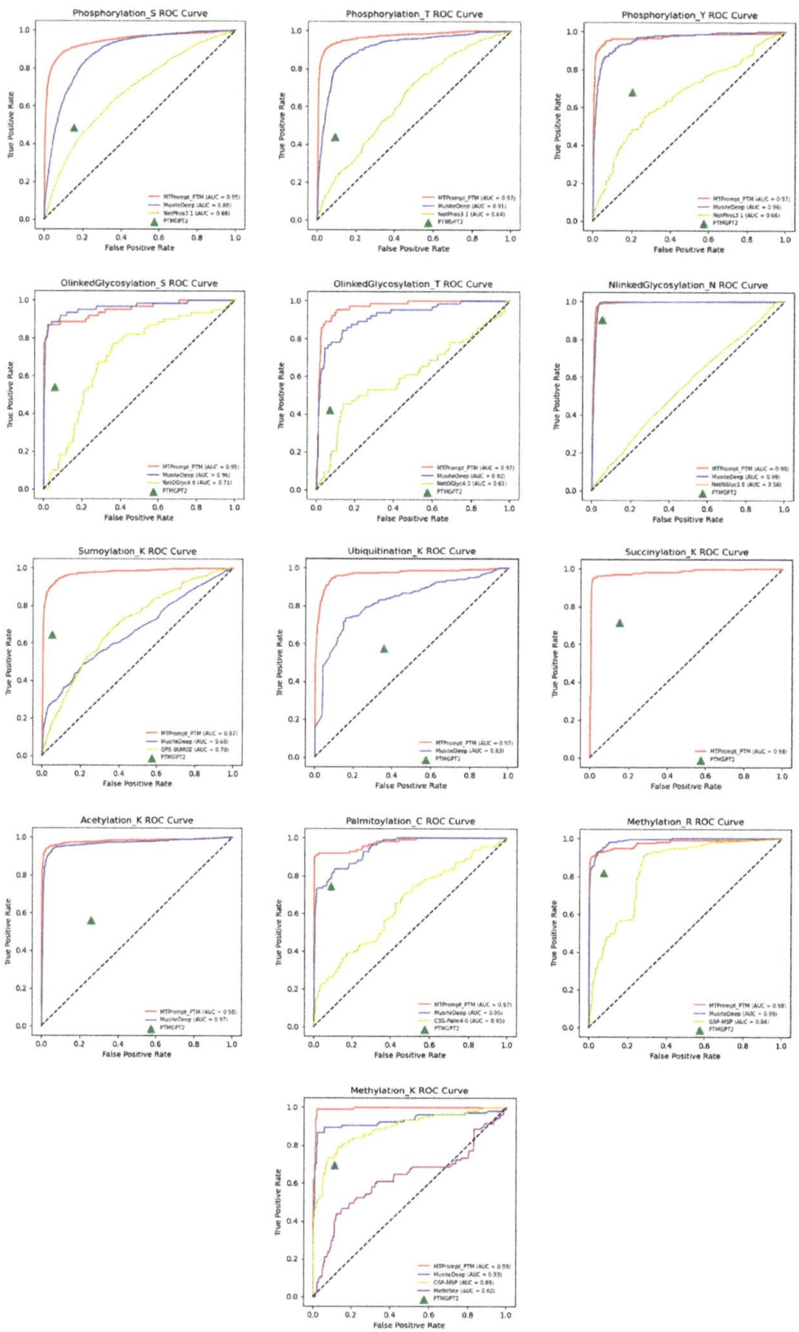

Figure 4. Performance comparison of AUC on 13 PTM types.

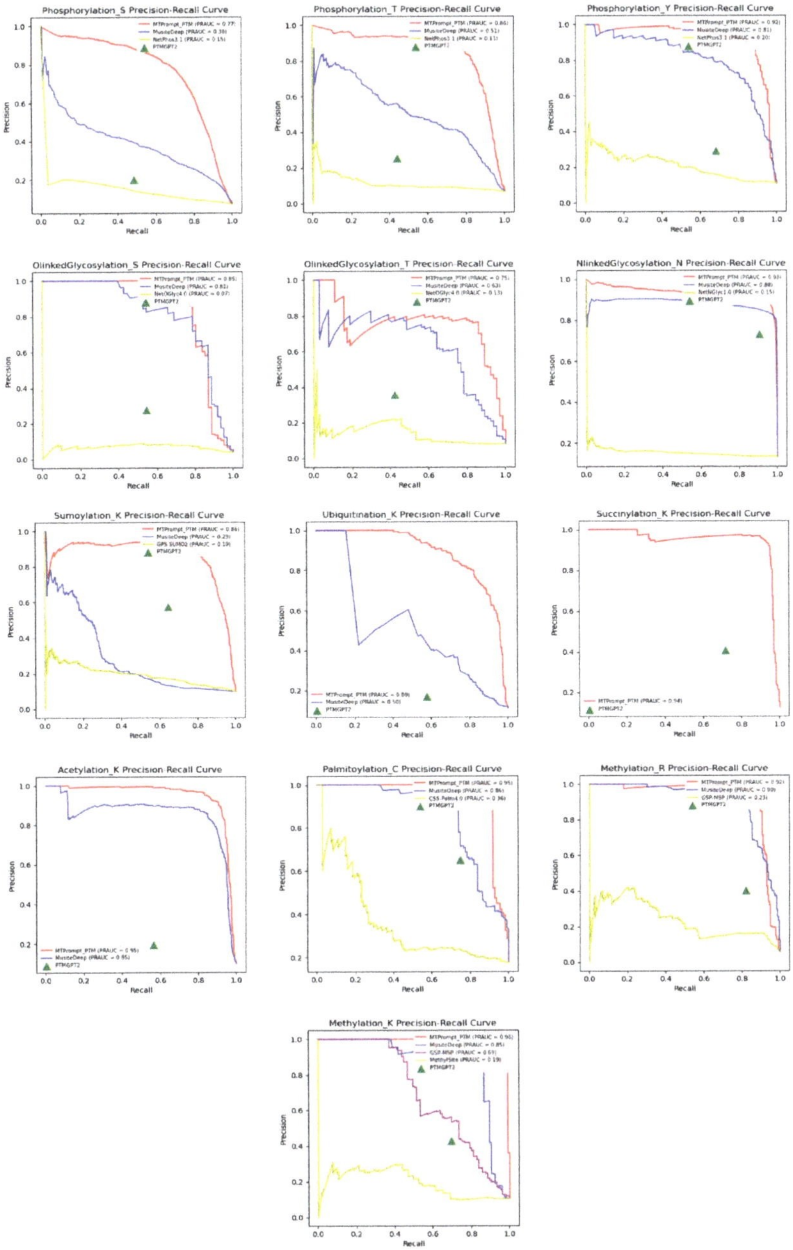

Figure 5. Performance comparison of PRAUC on 13 PTM types.

Table 4. Performance comparison of PTM prediction models across different kinase families.

Kinase Type	Number of Kinase Sites in Training Set	Method	Number of Kinase Sites in Testing Set	Number of Predicted Kinase Sites in Testing Set	Accuracy
AGC	879	MTPrompt-PTM	17	15	**0.882**
		MusiteDeep	17	14	0.824
		PTMGPT2	17	7	0.412
		NetPhos3.1	17	12	0.706
CAMK	392	MTPrompt-PTM	2	1	**0.5**
		MusiteDeep	2	2	1
		PTMGPT2	2	1	**0.5**
		NetPhos3.1	2	2	1
CK1	48	MTPrompt-PTM	1	1	1
		MusiteDeep	1	1	1
		PTMGPT2	1	0	0
		NetPhos3.1	1	1	1
CMGC	1739	MTPrompt-PTM	47	43	**0.915**
		MusiteDeep	47	42	0.894
		PTMGPT2	47	21	0.447
		NetPhos3.1	47	40	0.851
Other	415	MTPrompt-PTM	9	1	0.111
		MusiteDeep	9	3	**0.333**
		PTMGPT2	9	1	0.111
		NetPhos3.1	9	2	0.222
STE	174	MTPrompt-PTM	2	2	**1**
		MusiteDeep	2	2	**1**
		PTMGPT2	2	2	**1**
		NetPhos3.1	2	2	**1**
TK	753	MTPrompt-PTM	4	4	**1**
		MusiteDeep	4	4	**1**
		PTMGPT2	4	2	0.5
		NetPhos3.1	4	2	0.5

Note: Numbers in bold represent the highest values achieved for accuracy within a given kinase type.

Figure 6 illustrates the performance on test subsets with varying levels of sequence similarity to the training data, evaluated using the F1 score. The test subsets, containing protein sequences with no more than 60%, 70%, 80%, 90%, and 100% similarity to the training set, were generated using CD-HIT-2D. Across nearly all PTM types, MTPrompt-PTM consistently outperformed other methods regardless of the sequence similarity thresholds. However, for certain PTMs, we observe a noticeable decline in MTPrompt-PTM's performance as the sequence similarity decreases. This degradation can be attributed to several factors. Some PTM types may have fewer annotated examples or less diversity in the training set. As a result, the model may be overfit to specific sequence patterns and struggle to generalize to more dissimilar sequences. In addition, certain PTMs, such as phosphorylation (S) and SUMOylation (K), are known to be highly context-dependent, often influenced by local sequence motifs or secondary structure elements. When the sequence similarity drops, these subtle cues may no longer be preserved, making it more challenging for the model to accurately identify modification sites. Another contributing factor may be the imbalance in positive and negative samples across the different similarity subsets. As the similarity threshold decreases, the number of true PTM sites that remain in the test set may decline disproportionately, leading to a more severe class imbalance and potentially skewing the model's predictions. Despite these challenges, MTPrompt-PTM still maintains relatively strong performance across all similarity levels, underscoring its robustness compared to existing tools.

Figure 6. Performance on test subsets with different levels of sequence similarity to the training data, evaluated by F1. The protein sequences of testing data that had no more than 60%, 70%, 80%, 90%, and 100% similarity to the training data were generated by CD-HIT-2D.

Figure 7 compares the performance of MTPrompt-PTM with that of MusiteDeep, PTMGPT2, and PTM-Mamba on independent benchmark datasets for phosphorylation and non-histone acetylation. For phosphorylation, MTPrompt-PTM consistently achieves the best overall performance, particularly excelling in its accuracy, precision, F1 score, and MCC. This suggests that our model makes more reliable and balanced predictions with fewer false positives and the better discrimination of true modification sites. PTM-Mamba shows the highest recall, indicating its sensitivity in detecting true sites, but this comes at the cost of lower precision, leading to more false positives. In non-histone acetylation, MTPrompt-PTM again outperforms other methods in terms of accuracy and precision, demonstrating its ability to correctly identify modification sites with fewer errors. While MusiteDeep and PTM-Mamba have comparable recall values, their lower precision reduces their overall predictive quality. The differences may be due to MTPrompt-PTM's multi-task prompt tuning strategy and structure-aware embedding, which enhance its specificity and robustness across diverse PTM types. Overall, these results highlight that MTPrompt-PTM strikes a better balance between sensitivity and specificity, making it a more effective tool for PTM site prediction compared to competing methods.

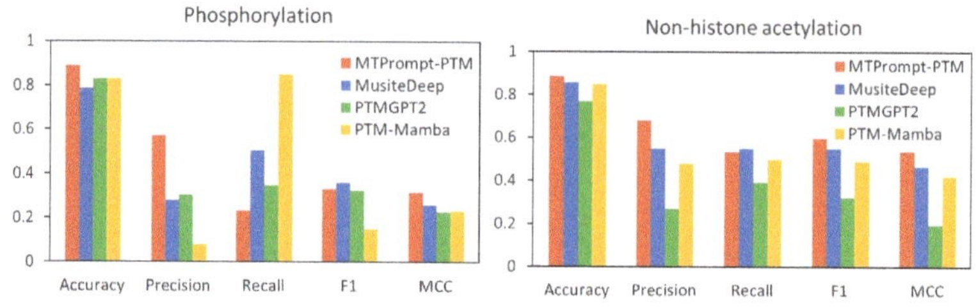

Figure 7. Comparative analysis of PTM prediction tools on independent phosphorylation and acetylation data.

3.2. Comparison with Single-Task Models on Different PTM Types

To evaluate the effectiveness of our proposed multi-task architecture, we compared it against 13 independently trained single-task models on our non-redundant dataset, with each model trained on a single PTM type. Both the multi-task and single-task models shared the same pre-trained backbone and utilized the same predefined task tokens, ensuring a fair comparison.

The results in Figure 8 indicate that phosphorylation (S, T, and Y) exhibited varying degrees of improvement in the multi-task model. Phosphorylation (S) showed only a 2% improvement in the AUPRC, likely due to its large training dataset, which may reduce its dependency on knowledge transfer from other PTM types. In contrast, phosphorylation (T) showed substantial gains, with threonine improving by 9% (AUPRC), suggesting that phosphorylation (T) benefits from shared knowledge with phosphorylation (S). O-linked glycosylation (S and T) demonstrated strong improvements across all metrics, with O-linked glycosylation (S) achieving improvements of 11.1% (AUPRC) and O-linked glycosylation (T) seeing the highest gains, with the AUPRC increasing by 11.2%, suggesting a possible interaction with phosphorylation. N-linked glycosylation (N) performed slightly worse in the multi-task setting, with an AUPRC decrease of 0.1%, indicating that asparagine may be more independent and less influenced by other PTM types. Acetylation (K), ubiquitination (K), succinylation (K), and SUMOylation (K) benefited from the multi-task model, showing improvements of 9.5%, 2.9%, 10.2%, and 2.5% (AUPRC), suggesting that PTMs occurring on lysine (K) may support each other through shared information. Overall, the results demonstrate that multi-task learning improves the performance for PTMs that share functional or structural similarities, such as phosphorylation and O-linked glycosylation. However, PTMs that are more independent, such as N-linked glycosylation (N), may not benefit as much from multi-task training. Additionally, PTMs on lysine (K) appear to influence each other, as seen in the gains for acetylation, ubiquitination, SUMOylation, and succinylation. The ROC curves are shown in Supplementary Figure S1. From the AUROC, we observe that the performance of the multi-task model is similar to that of the single-task model. This could be due to the imbalance between negative and positive samples. The improved AUPRC of the multi-task model likely arises from its ability to generalize across multiple PTM types, which helps to enhance its precision and recall for the positive class, especially in the presence of class imbalances. While the multi-task setup does not significantly impact the AUROC (as the AUROC is less sensitive to class imbalances), it has a more pronounced effect on the precision–recall performance.

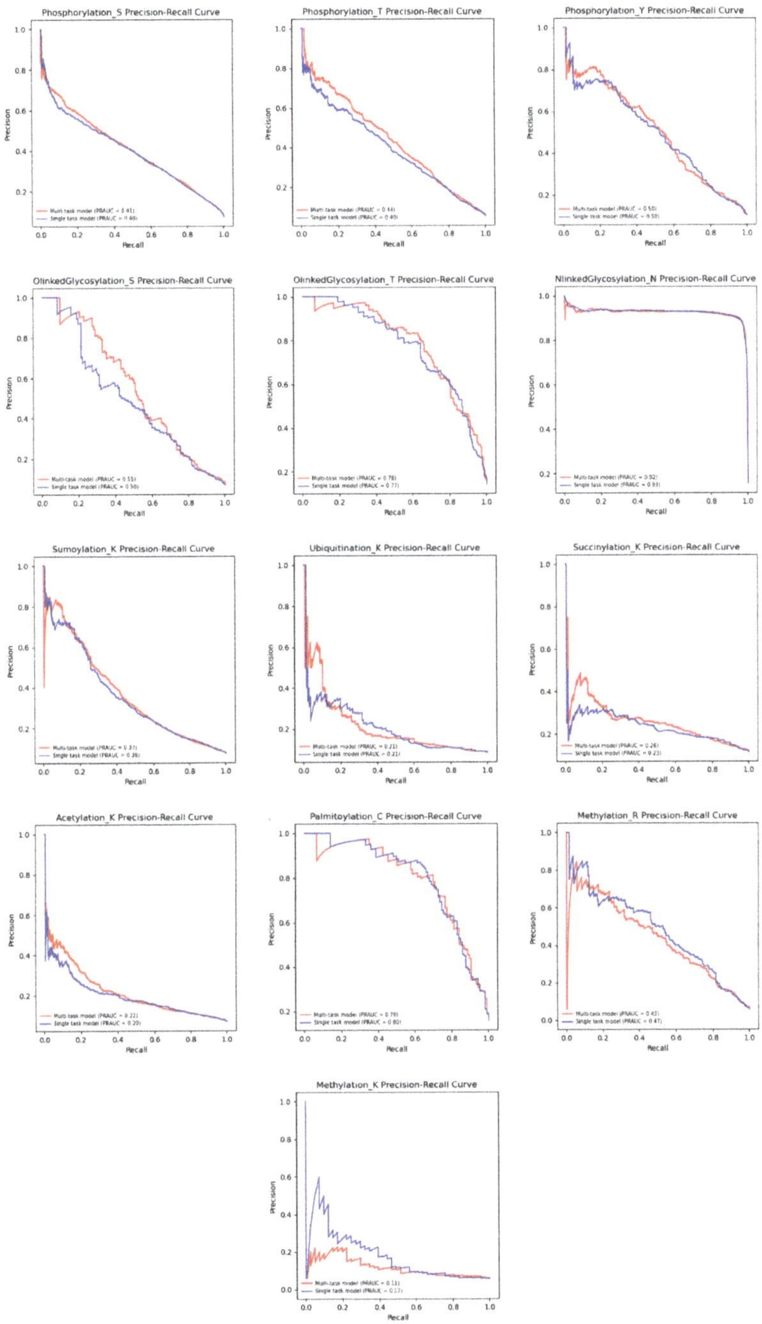

Figure 8. Performance comparison of PRAUC between multi-task and single-task models on 13 PTM types.

The F1 and MCC can show the improvements in the multi-task model as well, as presented in Table 5. From this table, we can see that MTPrompt-PTM achieves better performance than the single-task model in most PTM types. Notably, for phosphorylation

(T), phosphorylation (Y), O-linked glycosylation (S and T), and SUMOylation (K), the multi-task model yields higher F1 and MCC scores, indicating that leveraging shared knowledge across tasks enhances the prediction accuracy for these modification types. For example, in O-linked glycosylation (T), the F1 and MCC scores of the multi-task model reach 0.716 and 0.685, respectively, outperforming the single-task model (0.670/0.634). Although the single-task model performs slightly better in a few PTMs, such as N-linked glycosylation (N) and methylation (R), the difference is marginal, suggesting that the multi-task framework does not significantly compromise the performance even in well-characterized or distinct PTMs. Overall, the results demonstrate that the multi-task model provides more stable and generalized performance across diverse PTM types, especially those with limited training data or weaker individual signals.

Table 5. Performance comparison of F1 and MCC on MTPrompt-PTM and separately trained model.

	F1/MCC	
PTM Type (Residue)	**Multi-Task Model**	**Single-Task Model**
Phosphorylation (S)	0.428/**0.384**	**0.429**/0.383
Phosphorylation (T)	**0.461/0.432**	0.439/0.406
Phosphorylation (Y)	**0.503/0.459**	0.498/0.448
N-Linked Glycosylation (N)	0.918/0.902	**0.922/0.907**
O-Linked Glycosylation (S)	**0.524/0.5**	0.487/0.447
O-Linked Glycosylation (T)	**0.716/0.685**	0.670/0.634
Palmitoylation (C)	**0.74/0.697**	0.730/0.685
Acetylation (K)	**0.214/0.206**	0.189/0.180
Ubiquitination (K)	**0.081/0.129**	0.051/0.074
Succinylation (K)	**0.208/0.176**	0.144/0.109
SUMOylation (K)	0.352/**0.342**	**0.361**/0.332
Methylation (K)	0/0.142	**0.089/0.143**
Methylation (R)	0.431/0.414	**0.470/0.454**

Note: Numbers in bold represent the highest values achieved for F1 and MCC within a given PTM type.

3.3. Ablation Study

3.3.1. Comparison with Multi-Task Model Without Knowledge Distillation on Different PTM Types

To assess the impact of knowledge distillation, we compared MTPrompt-PTM with a baseline multi-task model trained solely on hard labels without distillation. As shown in Table 6, MTPrompt-PTM consistently achieved higher F1 and MCC scores across all PTM types, demonstrating improved predictive accuracy and robustness.

Table 7 presents the AUROC and AUPRC comparisons, where MTPrompt-PTM achieves notably higher AUPRC values in most cases. For instance, in O-linked glycosylation (S) and (T), the AUPRC increased from 0.518 and 0.761 to 0.552 and 0.784, respectively. This improvement in the AUPRC is particularly significant given the class imbalance typically present in PTM site prediction tasks.

Table 6. Performance comparison of F1 and MCC on MTPrompt-PTM and multi-task model without knowledge distillation.

PTM Type (Residue)	F1/MCC	
	Multi-Task Model with Knowledge Distillation	**Multi-Task Model Without Knowledge Distillation**
Phosphorylation (S)	**0.428/0.384**	0.341/0.338
Phosphorylation (T)	**0.461/0.432**	0.389/0.381
Phosphorylation (Y)	**0.503/0.459**	0.448/0.424
N-Linked Glycosylation (N)	**0.918/0.902**	0.916/0.901
O-Linked Glycosylation (S)	**0.524/0.5**	0.45/0.446
O-Linked Glycosylation (T)	**0.716/0.685**	0.667/0.634
Palmitoylation (C)	**0.74/0.697**	0.719/0.678
Acetylation (K)	**0.214/0.206**	0.160/0.164
Ubiquitination (K)	**0.081/0.129**	0.081/0.129
Succinylation (K)	**0.208/0.176**	0.044/0.062
SUMOylation (K)	**0.352/0.342**	0.253/0.301
Methylation (K)	**0/0.142**	0/0.142
Methylation (R)	**0.431**/0.414	0.411/**0.415**

Note: Numbers in bold represent the highest values achieved for F1 and MCC within a given PTM type.

Table 7. Performance comparison of AUROC and AUPRC of MTPrompt-PTM and multi-task model without knowledge distillation.

PTM Type	AUROC/AUPRC	
	Multi-Task Model with Knowledge Distillation	**Multi-Task Model Without Knowledge Distillation**
Phosphorylation (S)	**0.866**/0.409	0.866/**0.411**
Phosphorylation (T)	0.878/**0.436**	**0.879**/0.429
Phosphorylation (Y)	0.844/**0.504**	**0.845**/0.496
N-Linked Glycosylation (N)	0.990/0.924	**0.991/0.927**
O-Linked Glycosylation (S)	**0.870/0.552**	0.864/0.518
O-Linked Glycosylation (T)	**0.933/0.784**	0.933/0.761
Palmitoylation (C)	**0.929**/0.791	0.925/**0.792**
Acetylation (K)	0.739/**0.220**	**0.742**/0.212
Ubiquitination (K)	**0.669/0.215**	0.669/0.215
Succinylation (K)	0.720/0.260	**0.722/0.268**
SUMOylation (K)	**0.798/0.369**	0.797/0.359
Methylation (K)	0.663/0.122	**0.669/0.130**
Methylation (R)	0.893/0.438	**0.910/0.450**

Note: Numbers in bold represent the highest values achieved for AUROC and AUPRC within a given PTM type.

These results confirm that incorporating soft labels from single-task models during training enables the multi-task framework to capture richer probabilistic information and subtle interdependencies across PTM types. This additional knowledge improves its generalization, especially for low-signal or data-scarce PTMs, ultimately leading to better overall performance than training on hard labels alone.

3.3.2. Comparison with Fine-Tuning the Last Two Layers of S-PLM on Different PTM Types

To further evaluate the effectiveness of our prompt tuning strategy, we compared MTPrompt-PTM with a commonly used fine-tuning approach that updates only the last two layers of the pre-trained S-PLM model. This comparison assessed whether prompt tuning could match or exceed the performance while minimizing the computational overhead. As shown in Table 8, MTPrompt-PTM achieved higher F1 and MCC scores across nearly all PTM types. Notably, large gains are observed in phosphorylation (T) and O-linked glycosylation (T), where the F1/MCC scores increase from 0.403/0.392 and 0.548/0.554 to 0.461/0.432 and 0.716/0.685, respectively. These results indicate that prompt tuning significantly enhances both the precision and robustness in PTM site prediction. Although the fine-tuned model slightly outperforms MTPrompt-PTM in ubiquitination (K), methylation (K), and methylation (R), the performance gap is minimal.

Table 8. Performance comparison of F1 and MCC of MTPrompt-PTM and multi-task model with fine-tuning in the last two layers of S-PLM.

	F1/MCC	
PTM Type	**Multi-Task Model with Prompt Tuning**	**Multi-Task Model with Fine-Tuning in Last Two Layers of S-PLM v2**
Phosphorylation (S)	**0.428/0.384**	0.355/0.340
Phosphorylation (T)	**0.461/0.432**	0.403/0.392
Phosphorylation (Y)	**0.503/0.459**	0.450/0.427
N-Linked Glycosylation (N)	**0.918/0.902**	0.916/0.900
O-Linked Glycosylation (S)	**0.524/0.5**	0.49/0.481
O-Linked Glycosylation (T)	**0.716/0.685**	0.548/0.554
Palmitoylation (C)	**0.74/0.697**	0.695/0.651
Acetylation (K)	**0.214/0.206**	0.214/0.206
Ubiquitination (K)	0.081/0.129	**0.082/0.156**
Succinylation (K)	**0.208/0.176**	**0.11/0.124**
SUMOylation (K)	**0.352/0.342**	**0.268/0.304**
Methylation (K)	0/0.142	0.048/0.152
Methylation (R)	0.431/0.414	0.435/0.435

Note: Numbers in bold represent the highest values achieved for F1 and MCC within a given PTM type.

Table 9 further confirms this trend through AUROC and AUPRC comparisons. MTPrompt-PTM shows a notable advantage in the AUPRC, especially for PTMs like O-linked glycosylation (S) and O-linked glycosylation (T), with improvements from 0.525 and 0.757 to 0.552 and 0.784, respectively. These metrics are particularly important in highly imbalanced datasets, where the ability to correctly identify true positives is critical. Furthermore, in phosphorylation (Y), the AUPRC improves from 0.499 to 0.504, and, in SUMOylation (K), from 0.354 to 0.369, reinforcing the consistency of the performance gains across diverse PTM types.

This improvement may be attributed to two key factors. First, by keeping the entire ESM2 model frozen, MTPrompt-PTM retains the broad, general-purpose protein representations learned during large-scale pre-training, avoiding the risk of overfitting or forgetting.

Second, the integration of task-specific prompt embeddings enables fine-grained adaptation to each PTM type, capturing subtle biochemical cues that are often lost in shallow fine-tuning. In contrast, updating only the last two layers may be insufficient to extract the deep contextual information required for accurate PTM site identification.

Table 9. Performance comparison of AUROC and AUPRC of MTPrompt-PTM and multi-task model with fine-tuning in the last two layers of S-PLM.

	AUROC/AUPRC	
PTM Type	**Multi-Task Model with Prompt Tuning**	**Multi-Task Model with Fine-Tuning in Last Two Layers of S-PLM v2**
Phosphorylation (S)	**0.866/0.409**	0.865/0.409
Phosphorylation (T)	0.878/**0.436**	**0.882**/0.434
Phosphorylation (Y)	**0.844/0.504**	0.837/0.499
N-Linked Glycosylation (N)	0.990/0.924	**0.991/0.930**
O-Linked Glycosylation (S)	**0.870/0.552**	0.845/0.525
O-Linked Glycosylation (T)	**0.933/0.784**	0.922/0.757
Palmitoylation (C)	**0.929**/0.791	0.927/**0.795**
Acetylation (K)	**0.739/0.220**	0.739/0.220
Ubiquitination (K)	0.669/**0.215**	**0.672**/0.207
Succinylation (K)	**0.720/0.260**	0.717/0.259
SUMOylation (K)	**0.798/0.369**	0.792/0.354
Methylation (K)	0.663/0.122	**0.704/0.289**
Methylation (R)	0.893/0.438	**0.903/0.461**

Note: Numbers in bold represent the highest values achieved for AUROC and PRAUC within a given PTM type.

4. Discussion

Exposing a model to a diverse set of tasks can serve as an effective form of regularization, reducing the risk of overfitting by encouraging the learning of generalizable patterns rather than memorizing task-specific details. A key advantage of multi-task learning lies in its ability to facilitate knowledge transfer between related tasks, i.e., improvements in one task can enhance the performance in others. Building on this principle, we developed MTPrompt-PTM, the first multi-task PTM prediction model capable of predicting 13 types of post-translational modifications (PTMs): phosphorylation (S, T, Y), N-linked glycosylation (N), O-linked glycosylation (S, T), ubiquitination (K), acetylation (K), methylation (K, R), SUMOylation (K), succinylation (K), and palmitoylation (C). At inference time, users simply need to provide a protein sequence along with the PTM type(s) that they wish to predict, making MTPrompt-PTM both versatile and user-friendly.

Unlike conventional PLM-based methods, MTPrompt-PTM leverages multi-task prompt tuning on the pre-trained S-PLM model, allowing it to adapt to diverse PTM types by incorporating task-specific signals. A decoder architecture composed of shared and task-specific layers further enables the model to capture both general and PTM-specific representations during training. To enhance the performance and generalization, knowledge distillation is employed, transferring insights from multiple single-task teacher models into a unified multi-task student model. Through extensive comparisons with single-task models and several state-of-the-art PTM prediction tools, MTPrompt-PTM consistently outperforms alternative methods across all PTM types, affirming the effectiveness of multi-task learning within this domain.

The effectiveness of MTPrompt-PTM can be attributed to three key factors. First, MTPrompt-PTM leverages the S-PLM v2 backbone, which captures both local and global sequence and structural information, providing a strong foundation for PTM prediction. Second, it employs multi-task prompt tuning, a lightweight fine-tuning method that efficiently adapts the PLM to the nuances of multiple PTM types while retaining the general-purpose knowledge encoded in the protein language model. This approach enables PTM prediction without compromising the pre-trained model's integrity. Third, MTPrompt-PTM incorporates a multi-PTM training framework with a knowledge distillation strategy, facilitating shared learning across different PTM types. This strategy enhances the performance, particularly for PTMs with limited training data.

However, several limitations exist. First, due to the computational complexity of processing long sequences, MTPrompt-PTM can only accept protein sequences of up to 1022 residues. This limitation could restrict its applicability in real-world scenarios, where longer sequences are common. Second, to better simulate real-world conditions, we separated the training and testing sets based on timestamps. However, some similarities between the training and testing sets remained. As the sequence similarity between the training and testing sets decreases, the performance also decreases. This could be due to data leakage, where information from the testing set unintentionally influences the training process, potentially causing overfitting. As a result, the model becomes overly specialized in sequences present in the training set and struggles to generalize to novel sequences in the test set.

In the future, we aim to extend our framework to support continuous learning, enabling it to accommodate additional modifications as new data become available. Expanding the dataset and incorporating more diverse annotations will improve the generalizability. However, challenges related to ensuring that continuous learning does not interfere with the performance of previous models need to be addressed. Additionally, the imbalanced nature of PTM training data may lead to biased predictions toward overrepresented classes. Future work should explore techniques such as focal loss, class reweighting, or data augmentation to address this imbalance and improve the model's fairness and accuracy.

5. Conclusions

MTPrompt-PTM represents a step forward in the scalable, multi-task prediction of post-translational modification sites. Instead of relying on fragmented single-task approaches, it unifies 13 PTM types into one flexible framework, enabling users to make efficient, type-specific predictions from a single model. Its consistent performance across benchmark datasets, kinase-specific analyses, and external validation scenarios underscores its potential for broad application in bioinformatics. Looking ahead, MTPrompt-PTM provides a foundation for continuous, modular learning as PTM databases expand, supporting future developments in multi-label PTM annotation at the proteome scale.

Supplementary Materials: The following supporting information can be downloaded at: https://www.mdpi.com/article/10.3390/biom15060843/s1, Figure S1: Performance comparison of AUROC between multi-task and single-task on 13 PTM types.

Author Contributions: Conceptualization, D.W., Y.H. and D.X.; methodology, D.W., D.X., F.H. and Y.H.; software, Y.H.; formal analysis, D.W., F.H. and D.X.; data curation, Y.H.; writing—original draft preparation, Y.H.; writing—review and editing, D.W., F.H., D.X. and Q.S.; supervision, D.X.; project administration, D.X.; funding acquisition, D.X. and Q.S. All authors have read and agreed to the published version of the manuscript.

Funding: This research was funded by the National Institutes of Health (grant R35GM126985 to D.X.) and the National Institutes of Health (grant R01LM014510 to Q.S.).

Institutional Review Board Statement: Not applicable.

Informed Consent Statement: Not applicable.

Data Availability Statement: The source code, data, and trained model are available at GitHub (https://github.com/hanye311/MTPrompt-PTM/) (accessed on 6 June 2025).

Acknowledgments: The authors thank the anonymous reviewers for their valuable suggestions.

Conflicts of Interest: The authors declare no conflicts of interest.

References

1. Humphrey, S.J.; James, D.E.; Mann, M. Protein phosphorylation: A major switch mechanism for metabolic regulation. *Trends Endocrinol. Metab.* **2015**, *26*, 676–687. [CrossRef] [PubMed]
2. Vu, L.D.; Gevaert, K.; De Smet, I. Protein language: Post-translational modifications talking to each other. *Trends Plant Sci.* **2018**, *23*, 1068–1080. [CrossRef] [PubMed]
3. Deribe, Y.L.; Pawson, T.; Dikic, I. Post-translational modifications in signal integration. *Nat. Struct. Mol. Biol.* **2010**, *17*, 666–672. [CrossRef] [PubMed]
4. Khoury, G.A.; Baliban, R.C.; Floudas, C.A. Proteome-wide post-translational modification statistics: Frequency analysis and curation of the swiss-prot database. *Sci. Rep.* **2011**, *1*, 90. [CrossRef]
5. Zhu, H.; Bilgin, M.; Snyder, M. Proteomics. *Annu. Rev. Biochem.* **2003**, *72*, 783–812. [CrossRef]
6. Olsen, J.V.; Blagoev, B.; Gnad, F.; Macek, B.; Kumar, C.; Mortensen, P.; Mann, M. Global, in vivo, and site-specific phosphorylation dynamics in signaling networks. *Cell* **2006**, *127*, 635–648. [CrossRef]
7. Renart, J.; Reiser, J.; Stark, G.R. Transfer of proteins from gels to diazobenzyloxymethyl-paper and detection with antisera: A method for studying antibody specificity and antigen structure. *Proc. Natl. Acad. Sci. USA* **1979**, *76*, 3116–3120. [CrossRef]
8. Chen, Z.; Liu, X.; Li, F.; Li, C.; Zhang, X.; Liu, B.; Zhou, Y.; Song, J. Large-scale comparative assessment of computational predictors for lysine post-translational modification sites. *Brief. Bioinform.* **2020**, *21*, 2065–2076. [CrossRef]
9. Esmaili, F.; Pourmirzaei, M.; Ramazi, S.; Shojaeilangari, S.; Yavari, E. A Review of Machine Learning and Algorithmic Methods for Protein Phosphorylation Sites Prediction. *arXiv* **2022**, arXiv:2208.04311. [CrossRef]
10. Blom, N.; Gammeltoft, S.; Brunak, S. Sequence and structure-based prediction of eukaryotic protein phosphorylation sites. *J. Mol. Biol.* **1999**, *294*, 1351–1362. [CrossRef]
11. Gupta, R.; Brunak, S. Prediction of glycosylation across the human proteome and the correlation to protein function. *Pac. Symp. Biocomput.* **2002**, *7*, 310–322. [PubMed]
12. Deng, W.; Wang, Y.; Ma, L.; Zhang, Y.; Ullah, S.; Xue, Y. Computational prediction of methylation types of covalently modified lysine and arginine residues in proteins. *Brief. Bioinform.* **2017**, *18*, 647–658. [CrossRef] [PubMed]
13. Biggar, K.K.; Ruiz-Blanco, Y.B.; Charih, F.; Fang, Q.; Connolly, J.; Frensemier, K.; Adhikary, H.; Li, S.S.C.; Green, J.R. MethylSight: Taking a wider view of lysine methylation through computer-aided discovery to provide insight into the human methyl-lysine proteome. *bioRxiv* **2018**. bioRxiv:274688.
14. Ertelt, M.; Mulligan, V.K.; Maguire, J.B.; Lyskov, S.; Moretti, R.; Schiffner, T.; Meiler, J.; Schroeder, C.T. Combining machine learning with structure-based protein design to predict and engineer post-translational modifications of proteins. *PLoS Comput. Biol.* **2024**, *20*, e1011939. [CrossRef]
15. Wang, D.; Zeng, S.; Xu, C.; Qiu, W.; Liang, Y.; Joshi, T.; Xu, D. MusiteDeep: A deep-learning framework for general and kinase-specific phosphorylation site prediction. *Bioinformatics* **2017**, *33*, 3909–3916. [CrossRef]
16. Wang, D.; Liang, Y.; Xu, D. Capsule network for protein post-translational modification site prediction. *Bioinformatics* **2019**, *35*, 2386–2394. [CrossRef]
17. Wang, D.; Liu, D.; Yuchi, J.; He, F.; Jiang, Y.; Cai, S.; Li, J.; Xu, D. MusiteDeep: A deep-learning based webserver for protein post-translational modification site prediction and visualization. *Nucleic Acids Res.* **2020**, *48*, W140–W146. [CrossRef]
18. Gou, Y.; Liu, D.; Chen, M.; Wei, Y.; Huang, X.; Han, C.; Feng, Z.; Zhang, C.; Lu, T.; Peng, D.; et al. GPS-SUMO 2.0: An updated online service for the prediction of SUMOylation sites and SUMO-interacting motifs. *Nucleic Acids Res.* **2024**, *52*, W238–W247. [CrossRef]
19. Lin, Z.; Akin, H.; Rao, R.; Hie, B.; Zhu, Z.; Lu, W.; Smetanin, N.; Verkuil, R.; Kabeli, O.; Shmueli, Y.; et al. Evolutionary-scale prediction of atomic-level protein structure with a language model. *Science* **2023**, *379*, 1123–1130. [CrossRef]
20. Elnaggar, A.; Heinzinger, M.; Dallago, C.; Rihawi, G.; Wang, Y.; Jones, L.; Gibbs, T.; Feher, T.; Angerer, C.; Steinegger, M.; et al. ProtTrans: Towards cracking the language of life's code through self-supervised deep learning and high performance computing. *arXiv* **2020**, arXiv:2007.06225.
21. Pakhrin, S.C.; Pokharel, P.; Bhattarai, A.; Kc, D.B. LMNglyPred: Prediction of human N-linked glycosylation sites using embeddings from a pre-trained protein language model. *Glycobiology* **2023**, *33*, 411–420. [CrossRef] [PubMed]

22. Alkuhlani, A.; Gad, W.; Roushdy, M.; Voskoglou, M.G.; Salem, A.M. PTG-PLM: Predicting Post-Translational Glycosylation and Glycation Sites Using Protein Language Models and Deep Learning. *Axioms* **2022**, *11*, 469. [CrossRef]
23. Pokharel, S.; Pratyush, P.; Ismail, H.D.; Ma, J.; KC, D.B. Integrating Embeddings from Multiple Protein Language Models to Improve Protein O-GlcNAc Site Prediction. *Int. J. Mol. Sci.* **2023**, *24*, 16000. [CrossRef] [PubMed]
24. Shrestha, P.; Kandel, J.; Tayara, H.; Chong, K.T. Post-translational modification prediction via prompt-based fine-tuning of a GPT-2 model. *Nat. Commun.* **2024**, *15*, 6699. [CrossRef]
25. Peng, F.Z.; Wang, C.; Chen, T.; Schussheim, B.; Vincoff, S.; Chatterjee, P. PTM-Mamba: A PTM-aware protein language model with bidirectional gated Mamba blocks. *Nat. Methods* **2025**, *22*, 945–949. [CrossRef]
26. Bludau, I.; Willems, S.; Zeng, W.-F.; Strauss, M.T.; Hansen, F.M.; Tanzer, M.C.; Karayel, O.; Schulman, B.A.; Mann, M. The structural context of posttranslational modifications at a proteome-wide scale. *PLoS Biol.* **2022**, *20*, e3001636. [CrossRef]
27. Wang, D.; Abbas, U.L.; Shao, Q.; Chen, J.; Xu, D. S-PLM: Structure-aware Protein Language Model via Contrastive Learning between Sequence and Structure. *Adv. Sci.* **2023**, *12*, e2404212. [CrossRef]
28. Lester, B.; Al-Rfou, R.; Constant, N. The Power of Scale for Parameter-Efficient Prompt Tuning. *arXiv* **2021**, arXiv:2104.08691.
29. Li, W.; Godzik, A. Cd-hit: A fast program for clustering and comparing large sets of protein or nucleotide sequences. *Bioinformatics* **2006**, *22*, 1658–1659. [CrossRef]
30. Brandes, N.; Ofer, D.; Peleg, Y.; Rappoport, N.; Linial, M. ProteinBERT: A universal deep-learning model of protein sequence and function. *Bioinformatics* **2022**, *38*, 2102–2110. [CrossRef]
31. Hornbeck, P.V.; Kornhauser, J.M.; Tkachev, S.; Bin Zhang, B.; Skrzypek, E.; Murray, B.; Latham, V.; Sullivan, M. PhosphoSitePlus: A comprehensive resource for investigating the structure and function of experimentally determined post-translational modifications in man and mouse. *Nucleic Acids Res.* **2012**, *40*, D261–D270. [CrossRef] [PubMed]
32. Meng, L.; Chen, X.; Cheng, K.; Chen, N.; Zheng, Z.; Wang, F.; Sun, H.; Wong, K.-C. TransPTM: A transformer-based model for non-histone acetylation site prediction. *Brief. Bioinform.* **2024**, *25*, bbae219. [CrossRef] [PubMed]
33. Zhang, Y.; Qin, Y.; Pourmirzaei, M.; Shao, Q.; Wang, D.; Xu, D. Enhancing Structure-aware Protein Language Models with Efficient Fine-tuning for Various Protein Prediction Tasks. *bioRxiv* **2025**. bioRxiv:2025.04.23.650337.
34. Jing, B.; Eismann, S.; Suriana, P.; Townshend, R.J.L.; Dror, R. Learning from Protein Structure with Geometric Vector Perceptrons. In Proceedings of the International Conference on Learning Representations (ICLR), Virtual Event, Austria, 3–7 May 2021.
35. Clark, K.; Luong, M.-T.; Khandelwal, U.; Manning, C.D.; Le, Q.V. BAM! Born-Again Multi-Task Networks for Natural Language Understanding. In Proceedings of the 57th Annual Meeting of the Association for Computational Linguistics, Florence, Italy, 28 July–2 August 2019; Association for Computational Linguistics: Stroudsburg, PA, USA, 2019; pp. 5931–5937.
36. Steentoft, C.; Vakhrushev, S.Y.; Joshi, H.J.; Kong, Y.; Vester-Christensen, M.B.; Schjoldager, K.T.; Lavrsen, K.; Dabelsteen, S.; Pedersen, N.B.; Marcos-Silva, L.; et al. Precision mapping of the human O-GalNAc glycoproteome through SimpleCell technology. *EMBO J.* **2013**, *32*, 1478–1488. [CrossRef] [PubMed]
37. Zhou, F.; Xue, Y.; Yao, X.; Xu, Y. CSS-Palm: Palmitoylation site prediction with a clustering and scoring strategy (CSS). *Bioinformatics* **2006**, *22*, 894–896. [CrossRef]
38. Ren, J.; Wen, L.; Gao, X.; Jin, C.; Xue, Y.; Yao, X. CSS-Palm 2.0: An updated software for palmitoylation sites prediction. *Protein Eng. Des. Sel.* **2008**, *21*, 639–644. [CrossRef]

biomolecules

MDPI

Article

Phi-Value and NMR Structural Analysis of a Coupled Native-State Prolyl Isomerization and Conformational Protein Folding Process

Ulrich Weininger [1], Maximilian von Delbrück [2], Franz X. Schmid [2] and Roman P. Jakob [3,*]

[1] Institute of Physics, Biophysics, Martin-Luther-University Halle-Wittenberg, 06120 Halle (Saale), Germany; ulrich.weininger@physik.uni-halle.de

[2] Laboratorium für Biochemie und Bayreuther Zentrum für Molekulare Biowissenschaften, Universität Bayreuth, 95447 Bayreuth, Germany; delbrueck@knauer.net (M.v.D.); fx.schmid@uni-bayreuth.de (F.X.S.)

[3] Focal Area Structural Biology, Biozentrum, University of Basel, Spitalstrasse 41, 4056 Basel, Switzerland

* Correspondence: roman.jakob@unibas.ch; Tel.: +41-61-267-2103; Fax: +41-61-267-2109

Abstract: Prolyl *cis/trans* isomerization is a rate-limiting step in protein folding, often coupling directly to the acquisition of native structure. Here, we investigated the interplay between folding and prolyl isomerization in the N2 domain of the gene-3-protein from filamentous phage fd, which adopts a native-state *cis/trans* equilibrium at Pro161. Using mutational and Φ-value analysis, we identified a discrete folding nucleus encompassing the β-strands surrounding Pro161. These native-like interactions form early in the folding pathway and provide the energy to shift the *cis/trans* equilibrium toward the *cis* form. Variations distant from the Pro161-loop have minimal impact on the *cis/trans* ratio, underscoring the spatial specificity and localized control of the isomerization process. Using NMR spectroscopy, we determined the structures for both native N2 forms. The *cis*- and *trans*-Pro161 conformations are overall identical and exhibit only slight differences around the Pro161-loop. The *cis*-conformation adopts a more compact structure with improved backbone hydrogen bonding, explaining the approximately 10 kJ·mol^{-1} stability increase of the *cis* state. Our findings highlight that prolyl isomerization in the N2 domain is governed by a localized folding nucleus rather than global stability changes. This localized energetic coupling ensures that proline isomerization is not simply a passive, slow step but an integral component of the folding landscape, optimizing both the formation of native structure and the establishment of the *cis*-conformation.

Keywords: protein folding; prolyl isomerization; NMR spectroscopy; Phi-value analysis; protein stability

1. Introduction

The *cis/trans* isomerization of peptidyl–prolyl bonds in proteins is inherently slow [1,2], with the *trans* form generally being favored in unfolded or newly synthesized polypeptide chains [3]. Proteins that contain *cis* prolyl bonds in their native state must undergo a *trans*-to-*cis* isomerization during folding. This isomerization process is closely linked to conformational folding, where initial folding often begins with certain proline residues in the incorrect (*trans*) state. These *trans*-proline residues create a kinetic barrier, temporarily halting folding and allowing conformational energy to accumulate [4–7]. This accumulated strain then drives the equilibrium toward the native *cis* form. After proline isomerization, folding can proceed rapidly to completion.

In the folded state, the *cis* isomer is favored because it allows for stronger stabilizing interactions compared to the *trans* isomer, resulting in an energetic coupling between folding and prolyl isomerization. The folded structure stabilizes the *cis* isomer, while the *cis* form further enhances the stability of the protein through additional interactions. During folding, this cooperative relationship evolves stepwise, as conformational energy initially drives the shift in the *cis/trans* equilibrium, but the stabilizing interactions of the native state only form after the slow *trans*-to-*cis* isomerization. This represents the rate-limiting step of the folding process and at the same time locks in the native state [8–10].

Prolyl isomerization not only serves as a critical rate-limiting step in protein folding but also functions as a molecular switch or timer in various biological processes [11–15]. Prolyl isomerases—enzymes that accelerate these isomerization reactions—are widely distributed and play essential roles in cellular function [16–18].

Studying folding intermediates or misfolded states containing non-native prolyl isomers remains challenging due to their instability and low population levels. These species are typically difficult to characterize structurally under equilibrium conditions, although, in rare cases, the coexistence of *cis* and *trans* isomers has been detected using NMR spectroscopy. However, the detailed structural characterization of these minor species has proven elusive [19–25].

The N2 domain of the gene-3-protein from the filamentous phage fd presents an ideal model system to investigate the link between folding and prolyl isomerization, as it enables the simultaneous assessment of the stability and unfolding/refolding kinetics of both the *cis* and *trans* forms [26]. The N2 domain folds independently, with Pro161, located at the tip of a β hairpin (Figure 1A), existing as a mixture of two native folded states differing in the *cis/trans* conformation of Pro161. During refolding, the *cis* content increases from 7% in the unfolded state to 90% in the folded state [26]. Both *cis* and *trans* forms of N2 represent true native states, displaying identical unfolding rates but differing in refolding kinetics. This difference suggests that the conformational energy driving *cis/trans* isomerization is already available in the folding transition state. The relationship between folding and prolyl isomerization in both the unfolded and folded states of N2 is effectively illustrated by the box model in Figure 1B, which links conformational folding with prolyl isomerization.

In our previous work, we used single- and double-mixing kinetic experiments, alongside mutational analysis, to determine the source of energy that shifts the *cis/trans* ratio of Pro161. We found that this energy largely originates from the two-stranded β sheet at the base of the Pro161 hairpin [27].

Building on this foundation, the current study expands the mutational analysis to 35 additional residues across the N2 domain. We use Φ-value analysis [28–33] to investigate the effects of each mutation on folding kinetics and stability, thereby gaining insights into the folding pathway of the N2 domain. Additionally, we used NMR spectroscopy to determine the solution structures of the *cis* and *trans* forms of Pro161, elucidating the molecular basis for their differing stabilities.

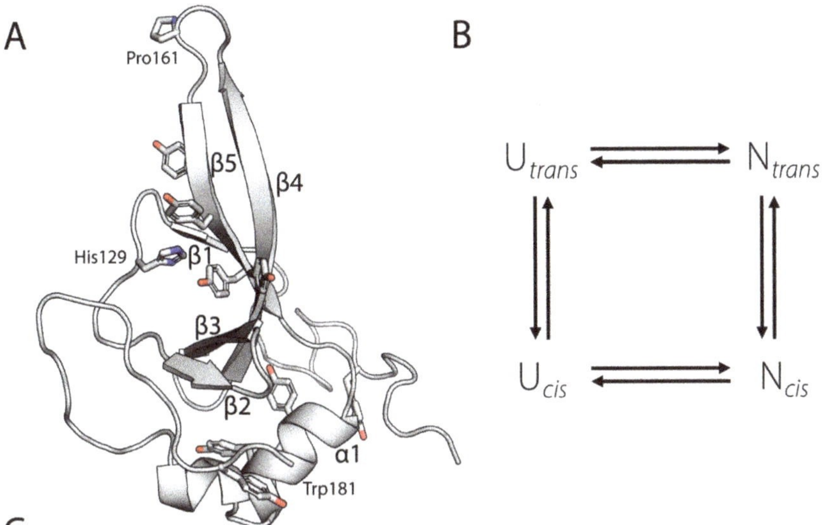

C

102 110 120 130 140 150
YINPLDGTYPPGTEQNPANPNPSLEESHPLNTFMFQNNRFRNRQGALTVYTG

160 170 180 190 200 205
TVTQGTD**P**VKTYYQYTPVSSKAMYDAYWNGKFRDCAFHSGFNEDLFVCEYQG

Figure 1. (**A**) Tertiary structure of the N2 domain of G3P (residues 102–205). The side chains of His129, Pro161, Trp181, and the nine Tyr residues are shown in stick representation. Trp181 is located behind helix1. The figure was prepared using PyMol [34] and the crystal structure of full-length G3P (PDB ID: 1G3P) [35]. (**B**) Model for the coupling between folding and prolyl isomerization of the N2 domain. (**C**) Amino acid sequence of the N2 domain. Pro161 is shown in bold, in blue are shown amino acid positions analyzed before [27], and amino acid positions analyzed in this work are colored green.

2. Materials and Methods

2.1. Mutagenesis, Protein Purification and Sample Preparation

The isolated N2 domain [residues 102–205 of the gene-3-protein of phage fd, extended by (His)6] with the stabilizing mutation Q129H was used as the reference (pseudo-wild-type) protein. The site-directed mutagenesis of N2′ was performed by BluntEnd-PCR [36] based on the expression plasmid pET11a (Novagen, Madison, WI, USA) The proteins were overproduced as inclusion bodies in *E. coli* BL21(DE3) and purified as described previously [26]. The concentrations of the N2-variants were determined via the absorption and the molar extinction coefficient ε_{280} = 19,000 $M^{-1}cm^{-1}$ at 280 nm. All buffers for spectroscopic measurements were dust filtered through 0.22 µm nylon filters before use and degassed in the desiccator with a membrane vacuum pump. The protein stock solutions were thawed at 4 °C, and any aggregates present were removed by centrifugation (approx. 30 min, 4 °C, 13,000 rpm, laboratory centrifuge).

2.2. Measurement of Near-UV Far-UV CD Spectra

All Circular dichroism spectra were measured using a Jasco J-600 (Tokyo, Japan) spectropolarimeter. Near-UV circular dichroism spectra (260–320 nm) were recorded with a protein concentration of 50 µM in 100 mM potassium phosphate, pH 7.0, in temperature-controlled 10 mm cuvettes. The step size was 0.2 nm at a measuring speed of 100 nm/min, the bandwidth was 2 nm, and the attenuation was 2 s. Spectra were measured ten times, averaged, and corrected for the contribution of the buffer. Far-UV spectra (185–250 nm) were measured in 10 mM potassium phosphate, pH 7.0, and a protein concentration of 5 μM.

2.3. Equilibrium Unfolding Transitions

For urea-induced unfolding, protein samples (1.0 μM) were incubated for 1 h at 15 °C in 100 mM K-phosphate, pH 7.0, and different concentrations of urea. The sample fluorescence was measured in 1 cm cuvettes at 340 nm (10 nm bandwidth) after excitation at 280 nm (5 nm bandwidth) (Hitachi F4010 fluorescence spectrometer). The data were analyzed according to a two-state model by assuming that ΔG_D as well as the fluorescence emissions of the folded and unfolded forms depend linearly on the urea concentration. A nonlinear least squares fit with proportional weighting of the experimental data was used to obtain ΔG_D as a function of urea concentration [37]. The heat-induced unfolding transitions were recorded in a Jasco J-600 spectropolarimeter equipped with a PTC 348 WI Peltier element at a protein concentration of 4 μM in 100 mM K-phosphate, pH 7.0 at a heating rate of 1 °C/min. The transitions were monitored by the CD signal increase at 222 nm with 1 nm bandwidth and 10 mm path length. The experimental data were analyzed based on the two-state approximation, with a heat capacity change ΔCp of 6400 J mol^{-1} K^{-1} [38].

2.4. Kinetic Experiments

All urea-induced unfolding and refolding experiments were performed in 100 mM K phosphate, pH 7.0, at 15 °C at a final protein concentration of 0.5 μM using a DX.17MV stopped-flow spectrometer from Applied Photophysics (Leatherhead, UK). The native or unfolded (in 4.4 M urea) protein was diluted 11-fold with urea solutions of different concentrations. The kinetics were monitored by the change in fluorescence over 320 nm after excitation at 280 nm (10 nm bandwidth) in an observation cell with a 2 mm path length. A 0.5 cm cell containing acetone was placed between the observation chamber and the photomultiplier to absorb scattered light from the excitation beam. The kinetics were measured at least eight times under identical conditions and averaged to improve the signal-to-noise ratio. In the analysis of the unfolding and refolding kinetics of the individual variants, we assumed that the folding kinetics of the *cis* and the *trans* forms are kinetically isolated by the slow U_t/U_c and N_t/N_c isomerizations and that the logarithms of the microscopic rate constants of unfolding and refolding depend linearly on the urea concentration. The ΔG_{UN} values were determined from the ratio of the rate constants for refolding and unfolding [$\Delta G_{UN} = -RT\ln(k_{UN}/k_{NU})$]. $\Delta\Delta G_{UN}$ is the difference between the ΔG_{UN} values of the mutant and the wild-type protein. The $\Delta\Delta G_{UN}^{\ddagger}$ values were derived from the ratio of the refolding rate constants of the mutant (mt) and the wild-type protein (wt) [$\Delta\Delta G_{UN}^{\ddagger} = -RT\ln(k_{UN}(mt)/k_{UN}(wt))$], and Φ is the $\Delta\Delta G_{UN}^{\ddagger}/\Delta\Delta G_{UN}$ ratio [39]. In order to keep the error caused by extrapolation as low as possible, the Φ-values for the unfolding were calculated for 2 M urea.

Interrupted unfolding experiments using double-mixing stopped-flow techniques were conducted to determine the *cis/trans* ratio in the folded forms of the N2′ variants at 15 °C. In the first step, 33 μM of folded N2′ protein (in 100 mM K phosphate, pH 7.0) was diluted 11-fold with 100 mM glycine buffer, initiating unfolding at a final pH of 2.0. Under these conditions, complete conformational unfolding occurred within 10 ms, while Pro161 *cis/trans* equilibration exhibited a time constant of 55 s [26]. After 100 ms, the N2′ variants were fully unfolded, but the *cis/trans* ratio remained virtually identical to that in the folded protein. Refolding was triggered after a 100 ms delay by an additional six-fold dilution, resulting in a final protein concentration of 0.5 μM in 100 mM K phosphate, pH 7.0. The *cis* content in the folded protein was determined by calculating the amplitude ratio of refolding reactions corresponding to the *cis* and *trans* isomers, considering slight fluorescence differences between N_c and N_t. Each experiment was repeated 10 times for all variants. Individual measurements were analyzed separately, with *cis* content variability remaining within 3%.

The Gibbs free energy that is necessary to shift the *cis*/*trans* equilibrium, referred to as the proline shift energy, is given by the equation $-RT\ln(K_N/K_U)$. Here, K_N and K_U represent the measured equilibrium constants for Pro161 *cis*/*trans* isomerization in the unfolded state ($K_U = [U_t]/[U_c]$) and in the native state ($K_N = [N_t]/[N_c]$), respectively.

2.5. NMR Spectroscopy

All NMR experiments were performed on a Bruker Avance II 600 spectrometer at 15 °C in 100 mM K phosphate buffer and 10% (*v/v*) D_2O at pH 7.0. Spectra were processed using NMRPipe [40] and analyzed using NMRView [41]. Backbone resonances have been assigned previously [42]. Aliphatic side chains were assigned by H(C)CH-TOCSY [43], and side chain NH and aromatic side chains by NOEs. NOEs for the structure determination were derived from 3D-NOESY-HSQC experiments for ^{15}N and ^{13}C aliphatic nuclei and a 2D NOESY experiment. Phi–Psi dihedral angle constraints were derived using TALOS [44]. H-bonds were introduced if all the following criteria were fulfilled: amide exchange of the corresponding amide is slowed down [42], NOE patterns are in agreement with H-bonds, they are located in secondary structure elements confirmed by chemical shifts and initial structures. $^{1}H/^{15}N$ RDCs were determined in 18 mg/mL PF1 phages from PROFOS. RDCs were used for isolated amide signals with $^{1}H^{15}N$ NOEs > 0.6. RDCs are located all over the structure. Structure calculations were performed using ARIA 2.3 [45]. 50 structures have been calculated using NOEs as ambiguous distance restraints, above mentioned Phi–Psi dihedral angle constraints, H-bonds and RDCs, and standard ARIA parameters for each Pro161 in *cis* or *trans*. Pro161 has been confirmed as *cis* in the NMR spectra; its surrounding was determined as flexible by $^{1}H^{15}N$ NOE experiments [42]. The *trans* structure was calculated with the same data set.

3. Results

3.1. Design of the Protein Variants

We employed protein engineering and Φ-value analysis [30,31,46] to investigate the transition state of folding for the N2 domain of the phage gene-3-protein. A total of 35 single-point mutations were introduced into the protein, and for each mutant, we measured the differences in Gibbs free energy of folding ($\Delta\Delta G_{UN}$), mutant minus wild-type protein, and activation free energy of refolding ($\Delta\Delta G_{UN}{}^{\ddagger}$). These values were derived from equilibrium unfolding transitions and folding kinetics, respectively. The Φ-value, calculated as the ratio of these two free energies, was determined for both the *trans* and *cis* forms of Pro161:

$$\Phi = \Delta\Delta G_{UN}{}^{\ddagger}/\Delta\Delta G_{UN}$$

A Φ-value of 1 indicates that the mutated side-chain is in a native-like environment in the transition state, contributing equally to the stability of both the native state and the transition state. In such cases, only the refolding kinetics are affected. Conversely, a Φ-value of 0 signifies that the side-chain is in an unfolded-like region of the transition state, affecting only the unfolding kinetics. Fractional Φ-values suggest partial structural formation at the mutation site or the presence of multiple folding pathways.

The N2 domain of the phage fd gene-3-protein, comprising residues 102–205 (Figure 1A), is only marginally stable in isolation. Its stability is enhanced by approximately 8 kJ·mol^{-1} in Gibbs free energy of denaturation (ΔG_D) by the Q129H mutation, identified via an in vitro selection [47]. For this study, we used the Q129H variant as our pseudo-wild-type protein, referred to as N2′.

Single mutations selected for the Φ-value analysis were evenly distributed over the protein (Figure 1C), except for regions previously analyzed (aa125–129; aa 136–139; aa 156–165,) [27]. With the exception of L198P, which was identified previously through in vitro selection [48], all residues were substituted by alanine. All N2′-variants were expressed in *E. coli* BL21 as inclusion bodies, refolded, and purified via Ni-affinity and size-exclusion chromatographies. Most variants exhibited lower refolding yields compared to the wild-type protein (Figure 2A). Five variants (F134A, F141A, Y168A, V171A, Y177A) aggregated after refolding, and two (V150A, Y151A) yielded only 1 mg. These seven residues are hydrophobic and located in the N2′ core, where their interactions are critical for protein stability (Figure 2B).

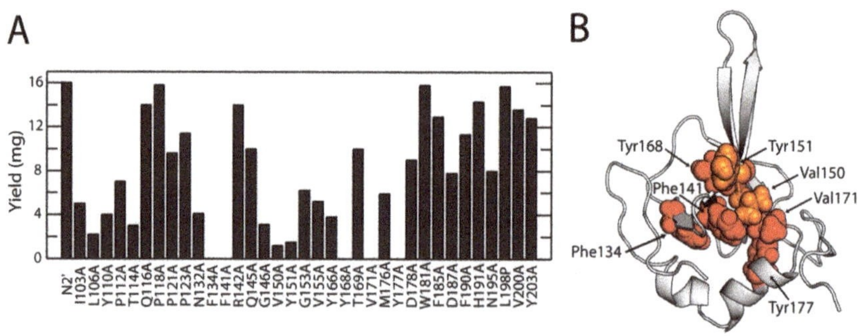

Figure 2. N2′ variant production (**A**) Yields (mg) of the individual N2′ variants per 2L fermentation. The 35 N2′ variants with the corresponding substitutions are given. (**B**) Protein variants with no or very low protein yields. When these amino acids were replaced with alanine, the respective variants gave very low protein yields (Val150, Tyr151, shown in orange) or precipitated (Phe134, Phe141, Tyr168, Val171, Tyr177, shown in red). These amino acids are part of the hydrophobic core of the N2′ domain and essential for stability.

CD spectroscopy was used as a quick and effective method for analyzing the secondary and tertiary structure of N2′ variants. Checking their correct folding is particularly important for Φ-value analysis, which is based on data from highly destabilized variants.

Far-UV CD spectra (185–250 nm) of the N2′ domain, consisting of five β-strands and a short α-helix, displays a β-sheet characteristic spectrum with a maximum at 194 nm and a minimum at 218 nm (Supplementary Figure S1A). The positive signal between 190 nm and 200 nm confirms the folded state of N2′ variants. The virtual identity of the far-UV CD spectra of N2′ and the variants (P112A, N132A, M176A, D187A, V200A) demonstrates that the substitutions to alanine did not affect the backbone structure of the folded variants.

The near-UV CD spectrum (260–320 nm) reflects the asymmetric immobilization of aromatic groups and thus the tertiary structure. N2′ shows a structural "fingerprint" with a maximum at 284 nm and a minimum at 266 nm (Supplementary Figure S1B), and again the variant proteins show the same spectra as N2′. This confirms that the substitutions to Ala also did not affect the tertiary structure of N2′.

3.2. Stabilities of the Protein Variants

The effect of alanine substitutions on the conformational stability of the N2′ domain was investigated by following thermal unfolding transitions and urea-induced equilibrium transitions. Thermal unfolding was monitored by measuring the ellipticity at 222 nm (Supplementary Figure S1C) and analyzed using a two-state model. To ensure comparability, the transitions were normalized to the fraction of native protein. Supplementary

Figure S2 summarizes the normalized thermally induced transitions for all N2′ variants, grouped by increasing stability (A to F). Gibbs free energies of thermal unfolding (ΔG_D) were extrapolated to 32 °C, the average midpoint temperature of all N2′ variants, to minimize extrapolation errors. Free energies were also calculated at 15 °C to correlate with thermodynamic stabilities derived from urea-induced transitions (Supplementary Table S1). For the highly destabilized variants L106A and V150A, the poorly defined baseline slopes of the native protein were adjusted using data from more stable variants. The N2′ variants displayed a wide range of mutation effects on stability, with most variants showing stability changes within ±5 kJ·mol^{-1}.

Urea-induced unfolding transitions of N2′ variants were measured using tryptophan emission at 340 nm (excitation at 280 nm) (Supplementary Figure S1D). Most variants showed fluorescence properties similar to the wild-type, except N2′-Y110A and N2′-W181A. The N2′ domain contains nine tyrosines and one tryptophan (W181) (Figure 1A), facilitating Förster Resonance Energy Transfer (FRET). The Tyrosine mutation Y110A reduced fluorescence by 60%, replacing W181 with alanine eliminated FRET, causing tyrosine fluorescence to increase due to environmental changes. Data were analyzed with a two-state model and normalized to native protein fractions (Supplementary Figure S3). Analysis of highly destabilized variants required fixed baselines. Thermodynamic parameters (Supplementary Table S1) showed cooperativity across variants, similar to the wild-type protein. ΔG values at 15 °C in 2 M urea were extrapolated for Φ-value calculations.

Stabilities from thermal and urea-induced unfolding transitions correlate well, as shown by the alignment of midpoints $[\text{urea}]_M$ and T_M in Figure 3A. Figure 3B shows the impact of alanine substitutions on N2′ stability ($\Delta\Delta G_D$). Destabilizing mutations are distributed throughout the domain. Strong destabilization at the protein N-terminus is followed by a region with reduced destabilization (residues 116–145). This less destabilized region encompasses the loops that precede and extend from β-strand 2 and 3 (Figure 1A), and it is followed by another region with strong destabilizing mutations (residues 150–190). Even residues near the C-terminus (V200, Y203) make significant contributions to protein stability.

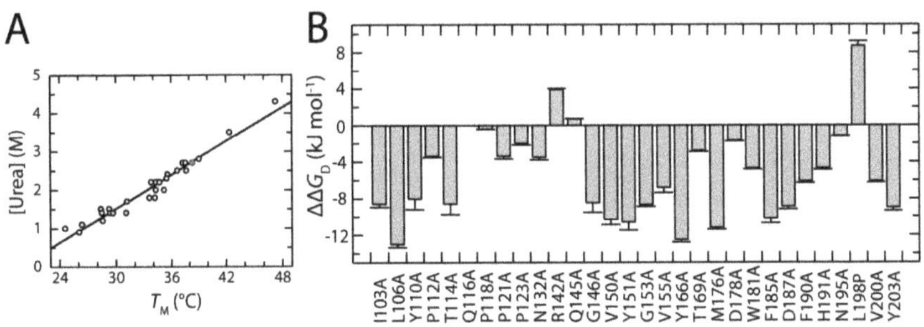

Figure 3. Stability analysis of the N2′ variants. (**A**) Comparison of the mid points of unfolding the N2′ variants (open dots) and the wild-type protein (filled red dot) from the thermally and urea-induced equilibrium transitions. The data shown are taken from Supplementary Table S1 and were evaluated using linear regression (filled line). (**B**) Free enthalpy difference $\Delta\Delta G_D{}^{15\,^\circ C}$ at 2 M urea of the individual N2′ variants compared to the wild-type protein.

3.3. Folding Kinetics of the Protein Variants

In stopped-flow experiments, the unfolding and refolding kinetics of the N2′ variants were measured by fluorescence as a function of the urea concentration. In the denatured state, N2′ exists as 93% *trans*- and 7% *cis*-Pro161 conformers [26], differing in stability and causing biexponential refolding kinetics. The main refolding phase (*trans*-conformer) has a large amplitude. The *cis*-conformer refolds more rapidly with a smaller amplitude. Slow proline isomerization (τ~100 s) is decoupled from folding and not analyzed further.

In contrast to biphasic refolding, N2′ unfolding follows a monoexponential course, as both conformers unfold at the same rate. Supplementary Figures S4 and S5 summarize the rates of all N2′ variants from refolding and unfolding experiments as a function of urea concentration, plotted in a semi-logarithmic fashion in so-called Chevron plots. For comparison with the wild-type protein, the fit to N2′ wild-type data (in red) is included in each Chevron plot.

The results of the Chevron analyses, based on a two-state model, are listed in Supplementary Table S2. All values are calculated at 2 M urea to minimize extrapolation errors. The unfolding arms of all N2′ variants were well-defined and analyzable. However, due to the shift of the Chevron plot to lower urea concentrations, the refolding arms of the *trans*-conformers of several strongly destabilized variants (I103A, L106A, Y110A, G146A, V150A, Y151A, G153A, M176A, F185A, D187A) could not be analyzed. The dependence of apparent rates on urea concentration is similar for the wild-type protein and N2′ variants, with nearly identical unfolding slopes producing comparable Chevron plot shapes. The largest differences occur in the unfolding arms, where substitutions often accelerate unfolding, and can also be seen in the small changes in free activation enthalpy ($\Delta\Delta G_{UN}{}^{\ddagger}$) for refolding (Figure 4A) and large enthalpy changes for unfolding ($\Delta\Delta G_{NU}{}^{\ddagger}$) (Figure 4B). The average β-Tanford value (β_T) of 0.66 for the N2′ variants indicates that approximately two-thirds of the hydrophobic interior surface is buried in the transition state already (Supplementary Table S2). The similar β_T values observed across all N2′ variants suggest that replacing individual amino acids did not alter the overall packing of the protein interior in the transition state.

The sum of $\Delta\Delta G_{UN}{}^{\ddagger}$ and $\Delta\Delta G_{NU}{}^{\ddagger}$ corresponds to the total destabilization caused by the amino acid substitution and indeed matches very well with the Gibbs free energy differences obtained from the equilibrium transitions, $\Delta\Delta G_D$, for most N2′ variants. Figure 4C illustrates this by plotting these $\Delta\Delta G_D$ values against the sum of the activation parameters, $\Delta\Delta G_{UN}{}^{\ddagger} + \Delta\Delta G_{NU}{}^{\ddagger}$, for the *cis*-conformer from the refolding and unfolding experiments. The slope of the regression line, close to one, indicates a very good correlation between the kinetic and thermodynamic measurements.

Figure 4D presents a summary of all available Φ-values for the *cis*-form of the N2′ domain from this work and Jakob & Schmid (2009) [27]. Our analysis focuses on the *cis* form because, for many strongly destabilized N2′ variants, analyzing the *trans* form was either infeasible or resulted in high errors. The figure highlights a highly uneven distribution of Φ-values. Most of the N2′ domain exhibits low Φ-values (shown in blue in Figure 4D), indicating that these amino acids form their stabilizing interactions only after passing the transition state.

Distinct from this general trend is the Pro161-loop, where high Φ-values reported by Jakob & Schmid (2009) are confirmed by elevated Φ-values in the previous and subsequent Beta-strand (e.g., V155A, Y166A). This region seems to establish native-like interactions very early during folding. Interestingly, two abrupt increases in Φ-values, unrelated to side-chain interactions, were caused by backbone mutations. In G146A and G153A, conformational restrictions due to the more rigid alanine likely hindered refolding, explaining the sharp increase in Φ.

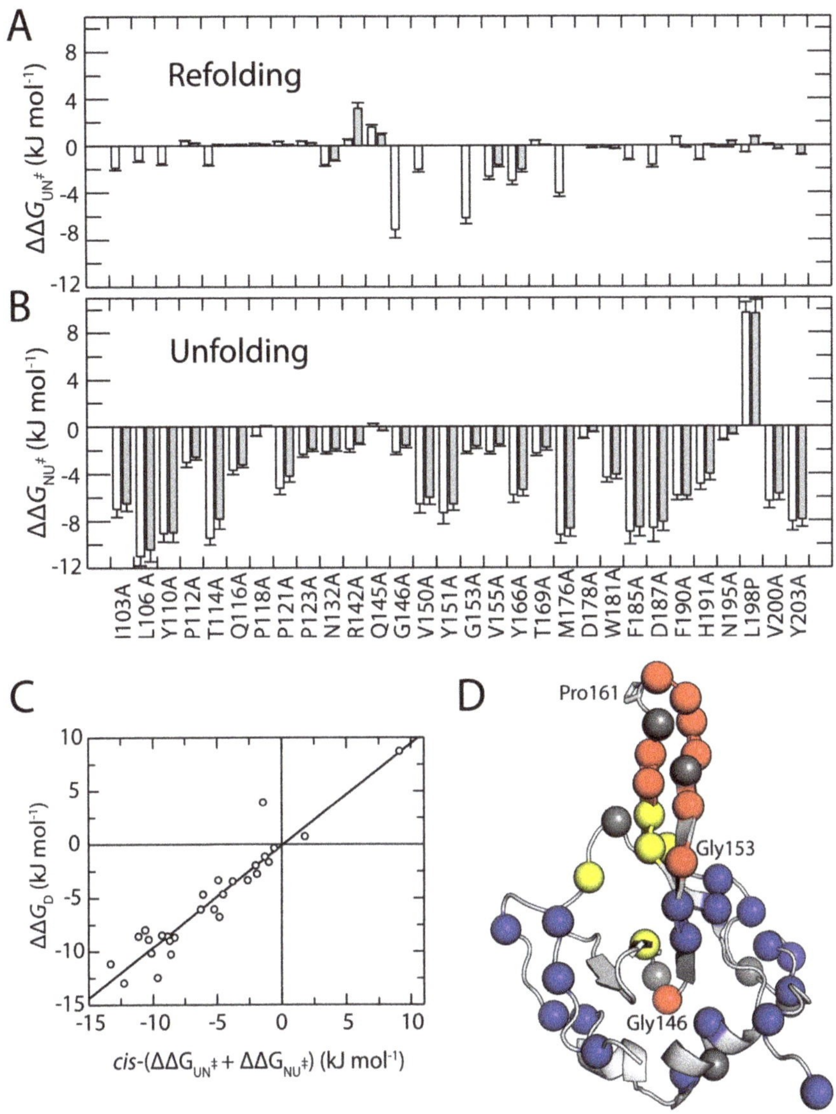

Figure 4. Comparison of the changes in the activation energy of refolding $\Delta\Delta G_{UN}{}^{\ddagger}$, (**A**) and activation energy of unfolding $\Delta\Delta G_{NU}{}^{\ddagger}$, (**B**) for the *trans* (gray) and the *cis* forms (white) of the variants. The $\Delta\Delta G_{NU}{}^{\ddagger}$ and $\Delta\Delta G_{UN}{}^{\ddagger}$ values are taken from Supplementary Table S2. The correlation of Gibbs free energy differences derived from kinetic measurements of the *cis*-conformer and equilibrium transitions is shown in (**C**). The regression line fitted to the data points follows the equation: $\Delta\Delta G_D = (\Delta\Delta G_{UN}{}^{\ddagger} + \Delta\Delta G_{NU}{}^{\ddagger}) \cdot 0.95$ kJ·mol^{-1} M^{-1}–0.07 kJ·mol^{-1} M^{-1}. All differences in Gibbs free energy are calculated for 2 M urea (at 15 °C). (**D**) The Cα-atom for amino acid positions analyzed within a Φ-value analysis are shown as spheres, including this work and previous analysis [27]. For simplification, only the Φ-values for the *cis*-Pro161 form are shown. The residues are color-coded according to their Φ-values: blue, $0.0 < \Phi < 0.3$; yellow, $0.3 < \Phi < 0.7$; red, $0.7 < \Phi < 1.0$; dark grey, residues with a $\Delta\Delta G_D < 2.0$ kJ·mol^{-1}.

These results demonstrate that the N2′ domain has a well-defined folding nucleus, primarily formed by the hairpin structure comprising β-strands 4 and 5 and the intervening loop around Pro161. Notably, this same region is responsible for the *cis/trans* equilibrium at Pro161.

3.4. Distant Variations Have No Impact on the cis/trans Ratio at Pro161

Pro161 is located in a large loop between two β-strands. Previous experiments suggested that the *cis/trans* ratio at Pro161 in the native protein is mainly determined by interactions between these β-strands. In the *cis*-conformation, these interactions are stronger, resulting in a more stable, native-like structure that favors the *cis*-conformer. In the presence of *trans*-Pro161, the connecting peptides are locally unfolded, structurally and energetically decoupling Pro161 from the β-strands (Jakob & Schmid, 2009). To measure both Uc → Nc and Ut → Nt refolding reactions with large amplitudes, a stopped-flow double-mixing protocol was employed.

In the first step, native N2′ was completely unfolded by a short 0.1 s long pH jump from 7.0 to 2.0 and then refolded in the initial buffer. Since the *cis/trans* ratio remained virtually unchanged during the 0.1 s unfolding pulse, the refolding amplitudes reflect the native *cis/trans* ratio at Pro161. The very small signal change from Nt ⇌ Nc equilibrium adjustment was taken into account when calculating the actual N_t content from the $U_t \rightarrow N_t$ amplitude [26]. The results from the double-mixing experiments clearly show that mutations outside the two β-strands have little impact on the equilibrium between the two conformers (Figure 5). The *cis*-fractions varied within a narrow range between 86% and 94%. The most significant reduction, down to 80%, was observed for the V155A substitution (Figure 5). V155A is located in the β-strand leading to the Pro161-loop (residues 157–164).

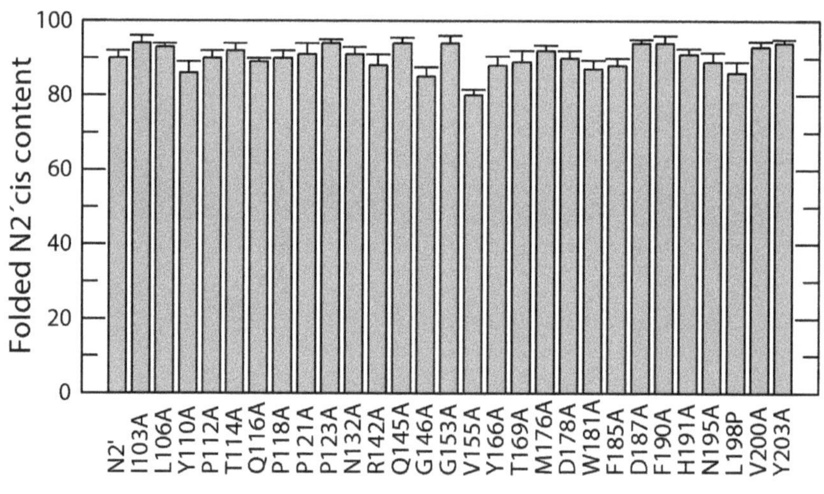

Figure 5. *Cis*-content at Pro161 in the folded N2′ variants. The data were determined in a double mixing experiment, as described in Materials and Methods, Section 2.4.

These double-mixing experiments demonstrate that global stability changes have no effect on the conformer ratio. Most studied positions are far from the exposed loop containing Pro161, supporting the idea that the *cis/trans* ratio is determined locally by interactions between the β-strands leading into the Pro161 loop and not by changes in the overall stability of N2′.

3.5. NMR Structure Determination of cis and trans-Form of N2′

Using Phi-value analysis, we found that the two β strands flanking Pro161 form the folding nucleus of the N2′ domain. This same region determines the native-state *cis*/*trans* ratio at Pro161. To obtain a structural model for both *cis*- and *trans*- forms of the N2 domain and explain their distinct stabilities, we used solution NMR spectroscopy. The N2′ backbone was previously assigned [42], and we have now assigned the side chain resonances as well. NMR spectra generally showed a single set of peaks (e.g., ^{1}H-^{15}N HSQC in Supplementary Figure S6A), indicating that we have just one main NMR state, with only a few weak additional signals, often from prolines themselves. These few weak signals suggest that, in the folded state, each proline primarily adopts either a single *cis* or *trans* conformation or, if a minor conformation exists, it has only a limited local impact on the structure.

Heteronuclear NOE (hNOE) measurements revealed that the Pro161-loop region exhibits notably low hNOE values [42]. Low hNOE values are indicative of elevated backbone flexibility and increased local mobility on the ps–ns timescale. This finding implies that, while the Pro161-loop forms the folding nucleus of the N2′ domain, it is less rigid and more dynamic than the neighbouring regions, suggesting that structural differences in the *trans* and *cis* forms of Pro are restricted to the β-sheets and loop directly nearby Pro161. In addition, differences in the β-sheets might be averaged out in the NMR spectra.

Of the nine prolines in N2′, NMR data clearly define only two states: Pro118 and Pro130 are in *trans*-state. For the NMR structure determination of both states, we treated Pro161 as *cis* or *trans* and all others as *trans*, based on available crystal structures [48]. Under these assumptions, we obtained a well-folded structural ensemble consistent with all NMR data (Supplementary Table S3). The structural ensembles are well-defined and only show deviations close to the protein termini and loop regions (Supplementary Figure S6B,C). With an r.s.m.d. of 1.5 Å, the NMR model of the isolated N2′ closely matches the X-ray structure (PDB ID: 1G3P) [35] of the N2 domain in full-length gene-3-protein (Figure 6A). Minor differences are restricted to loop regions, where low hNOE values confirm increased flexibility [42]. Thus, the isolated N2 domain exhibits a very similar fold as within the full-length protein. In addition, we calculated a structure for *trans*-Pro161. This structure uses the same NMR data as the *cis*-Pro161 structure, since we were unable to obtain any pure *trans*-Pro161 NMR data. The only difference is the actual conformation of Pro161. As expected by such an approach and considerations above, both *cis*-Pro161 and *trans*-Pro161 NMR structures are virtually identical, differing only slightly in the β-strands near Pro161 (Figure 6B). Both *cis* and *trans* models fit the NMR data equally well and exhibit nearly identical energies in the structural refinements (see Supplementary Table S3), indicating that structural differences are limited to the β-sheets and loops surrounding Pro161, and neither the absence of only *trans*-Pro161 NMR data nor the use of potentially only *cis*-Pro161 NMR data has a sizeable impact on the structures.

In the trans-Pro161 conformation, the Pro161-loop adopts a more extended structure (Figure 6C), resulting in four backbone hydrogen bonds forming between residues 155 and 165. In contrast, when Pro161 is *cis*, the loop becomes more compact. This compaction shortens the Gly158–Lys163 backbone hydrogen bond (3.0 Å in *cis* vs. 3.4 Å in *trans*) and introduces an additional backbone hydrogen bond (Thr159–Lys163) (Figure 6D). Together, these improved hydrogen-bonding interactions likely account for the approximately 10 kJ·mol^{-1} higher stability of the *cis* conformation.

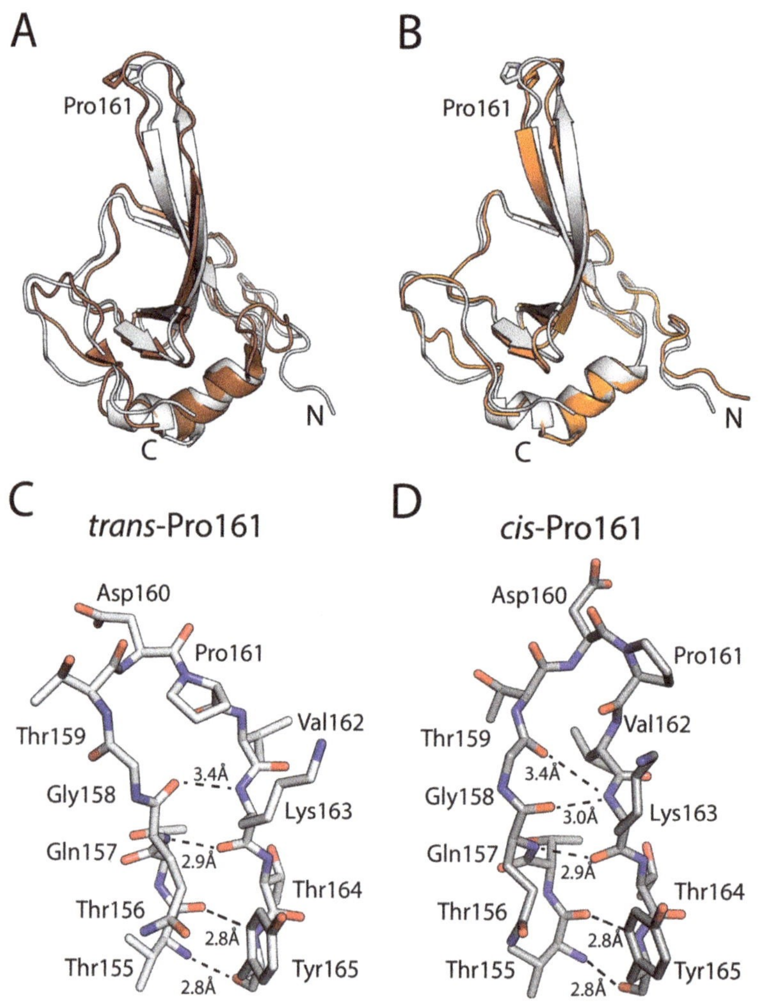

Figure 6. NMR structure determination of cis- and trans-conformation of N2′. (**A**) Superposition of the NMR model of N2′ containing *cis*-Pro161 (grey) with the crystal structure of the N2-domain within the gene-3-protein (PDB ID: 1G3P, brown). (**B**) Superposition of the NMR model of N2′ with cis-Pro161 (grey) and trans-Pro161 (orange). In (**A**,**B**) the N-and C-termini are indicated. Close-Up view comparison of (**C**) the *trans*-Pro (light grey, left) and (**D**) *cis*-Pro161 conformation (dark grey, right).

4. Discussion

Our results highlight the intricate balance between folding and prolyl isomerization in the N2 domain of the gene-3-protein and underscore the importance of localized interactions in determining the *cis/trans* equilibrium at Pro161. In the N2 domain, Pro161 is a critical residue where the *cis* isomer dominates the native state, despite the trans isomer being strongly favored in the unfolded ensemble [31,32]. Our expanded mutational analysis and subsequent Φ-value measurements reveal that the early formation of native-like interactions in the β-sheet region encompassing Pro161 is the central driver of this shift. These data support a model in which the energetic coupling between folding and prolyl isomerization is localized rather than global, allowing the N2 domain to accumulate the conformational energy required to flip the Pro161 peptide bond as folding progresses.

The Φ-value distribution (Figure 4D) shows that most residues form their native-like stabilizing interactions only after the folding transition state has been crossed, consistent with a cooperative, two-state folding mechanism. In contrast, the β-sheets surrounding Pro161 emerge as a clear "folding nucleus", as also found for other small proteins [28,49–55]. High Φ-values in the Pro161-loop region, as well as the heightened sensitivity to local substitutions (e.g., V155A) that alter the *cis*/*trans* ratio, demonstrate that these local β-strand interactions are established early and guide the protein toward the native *cis*-conformation. The lack of significant changes in the *cis*/*trans* ratio for mutations outside the immediate vicinity of Pro161 further reinforces the idea that the *cis*/*trans* equilibrium and its shift during folding is governed largely by local interactions (Figure 5). Alterations in stability or folding kinetics remote from the Pro161-loop do not substantially shift the *cis*/*trans* ratio. This points to a remarkable local spatial specificity for the mechanism of coupling between prolyl isomerization and conformational changes in the protein. Native-state prolyl isomerizations used as molecular switches are also controlled by precise local interactions to regulate communications pathways [15,16,56–60].

Our NMR structural studies corroborate these findings by demonstrating that the *cis* and *trans* conformations differ only slightly, primarily in backbone hydrogen bonding around Pro161. Both forms exhibit nearly identical global folds, implying no drastic structural rearrangements outside this local region (Figure 6). The more compact *cis* form of the Pro161-loop is stabilized by an additional hydrogen bond and a shortening of existing hydrogen bonds. Both factors contribute probably to the 10 $kJ \cdot mol^{-1}$ increase in stability. These subtle and localized changes confirm that the energetic and structural determinants of prolyl isomerization are localized to a narrowly confined region around Pro161. Thus, the NMR data link the mechanistic picture from kinetic and thermodynamic analyses to tangible structural features, illustrating how minimal structural rearrangements can yield significant energetic differences between *cis* and *trans* states.

In summary, our combined kinetic, thermodynamic, and structural results reveal that the *cis*/*trans* equilibrium at Pro161 is tightly intertwined with the early formation of native-like β-sheet interactions that shape the protein's folding landscape. This interplay ensures that prolyl isomerization is not a mere passive hurdle but an integral, controlled step within the folding pathway.

5. Conclusions

Mutational and Φ-value analyses pinpoint the β-strands surrounding Pro161 in the N2 domain as critical for both early protein folding and shifting the *cis*/trans equilibrium. Substitutions in distant regions have little effect on the *cis*/*trans* ratio of Pro161, underscoring the spatial specificity of this coupling. NMR structures of the *cis* and *trans* forms reveal nearly identical global folds, yet subtle differences in backbone hydrogen bonding around Pro161 provide an explanation for the higher thermodynamic stability of the *cis* conformation. Together, these results show that proline isomerization and the native-state Pro161 *cis*/*trans* equilibrium are integral parts of the protein folding mechanism, governed by local energetic and structural features in the protein.

Supplementary Materials: The following supporting information can be downloaded at: https://www.mdpi.com/article/10.3390/biom15020259/s1, Table S1: Stability data for N2′ and the variants; Table S2: Unfolding and refolding kinetics of N2′ variants; Table S3: Statistics of the NMR structure determination; Figure S1: Functional characterization and stability of the N2′ variants; Figure S2: Thermal induced unfolding transitions of N2′ variants; Figure S3: Urea induced equilibrium unfolding transitions; Figure S4: Folding kinetics; Figure S5: Chevron plots; Figure S6: NMR structural analysis.

Author Contributions: U.W., M.v.D., F.X.S. and R.P.J. designed the research, performed the research, and analyzed data; U.W. and R.P.J. wrote the paper. All authors have read and agreed to the published version of the manuscript.

Funding: This work was supported by the Fonds der Chemischen Industrie and the Deutsche Forschungsgemeinschaft—Project 186885367.

Institutional Review Board Statement: Not applicable.

Informed Consent Statement: Not applicable.

Data Availability Statement: Data are contained within the article and Supplementary Materials.

Acknowledgments: We thank all group members for helpful discussions.

Conflicts of Interest: The authors declare that the research was conducted in the absence of any commercial or financial relationships that could be construed as a potential conflict of interest.

Abbreviations

N2: N-terminal domain of gene-3-protein of phage fd; N2′, N2 containing the stabilizing mutation Q129H; T_M, midpoint of a thermal unfolding transition; ΔH_D, van't Hoff enthalpy of denaturation at T_M; [urea]$_M$, midpoint of a urea-induced unfolding transition; ΔG_D, Gibbs free energy of denaturation; m, cooperativity value of a denaturant (D)-induced equilibrium unfolding transition; k_{NU}, microscopic rate constant of unfolding; m_{NU}, kinetic m value of unfolding; k_{UN}, microscopic rate constant of refolding; m_{UN}, kinetic m value of refolding; β_T, Tanford value; U_c, U_t, N_c, N_t, unfolded and native forms of N2′ with a *cis*- or a *trans*-Pro161, respectively; Φ-value, ratio of the equilibrium and the activation Gibbs free energies of refolding; wt, the wild-type protein; mt, mutant.

References

1. Reimer, U.; Scherer, G.; Drewello, M.; Kruber, S.; Schutkowski, M.; Fischer, G. Side-chain effects on peptidyl-prolyl cis/trans isomerisation. *J. Mol. Biol.* **1998**, *279*, 449–460. [CrossRef] [PubMed]
2. Cheng, H.N.; Bovey, F.A. Cis-trans equilibrium and kinetic studies of acteyl-L-proline and glycyl-L-proline. *Biopolymers* **1977**, *16* 1465–1472. [CrossRef] [PubMed]
3. Brandts, J.F.; Halvorson, H.R.; Brennan, M. Consideration of the possibility that the slow step in protein denaturation reactions is due to cis-trans isomerism of proline residues. *Biochemistry* **1975**, *14*, 4953–4963. [CrossRef] [PubMed]
4. Pappenberger, G.; Bachmann, A.; Muller, R.; Aygun, H.; Engels, J.W.; Kiefhaber, T. Kinetic mechanism and catalysis of a native-state prolyl isomerization reaction. *J. Mol. Biol.* **2003**, *326*, 235–246. [CrossRef] [PubMed]
5. Lilie, H.; Rudolph, R.; Buchner, J. Association of antibody chains at different stages of folding: Prolyl isomerization occurs after formation of quaternary structure. *J. Mol. Biol.* **1995**, *248*, 190–201. [CrossRef]
6. Schreiber, G.; Fersht, A.R. The refolding of *cis*- and *trans*-peptidylprolyl isomers of Barstar. *Biochemistry* **1993**, *32*, 11195–11203 [CrossRef] [PubMed]
7. Wedemeyer, W.J.; Welker, E.; Scheraga, H.A. Proline cis-trans isomerization ansd protein folding. *Biochemistry* **2002**, *41*, 14637–14644. [CrossRef] [PubMed]
8. Aumuller, T.; Fischer, G. Bioactivity of folding intermediates studied by the recovery of enzymatic activity during refolding. *J. Mol. Biol.* **2008**, *376*, 1478–1492. [CrossRef]
9. Veeraraghavan, S.; Rodriguez-Gdiharpour, S.; MacKinnon, C.; Mcgee, W.A.; Pierce, M.M.; Nall, B.T. Prolyl isomerase as a probe of stability of slow-folding intermediates. *Biochemistry* **1995**, *34*, 12892–12902. [CrossRef]
10. Sakata, M.; Chatani, E.; Kameda, A.; Sakurai, K.; Naiki, H.; Goto, Y. Kinetic coupling of folding and prolyl isomerization of beta2-microglobulin studied by mutational analysis. *J. Mol. Biol.* **2008**, *382*, 1242–1255. [CrossRef] [PubMed]
11. Andreotti, A.H. Native state proline isomerization: An intrinsic molecular switch. *Biochemistry* **2003**, *42*, 9515–9524. [CrossRef]
12. Yaffe, M.B.; Schutkowski, M.; Shen, M.; Zhou, X.Z.; Stukenberg, P.T.; Rahfeld, J.U.; Xu, J.; Kuang, J.; Kirschner, M.W.; Fischer, G.; et al. Sequence-specific and phosphorylation-dependent proline isomerization: A potential mitotic regulatory mechanism. *Science* **1997**, *278*, 1957–1960. [CrossRef]

13. Schmidpeter, P.A.; Koch, J.R.; Schmid, F.X. Control of protein function by prolyl isomerization. *Biochim. Biophys. Acta* **2015**, *1850*, 1973–1982. [CrossRef] [PubMed]
14. Lu, K.P.; Zhou, X.Z. The prolyl isomerase PIN1: A pivotal new twist in phosphorylation signalling and disease. *Nat. Rev. Mol. Cell Biol.* **2007**, *8*, 904–916. [CrossRef] [PubMed]
15. Vogel, M.; Bukau, B.; Mayer, M.P. Allosteric regulation of Hsp70 chaperones by a proline switch. *Mol. Cell* **2006**, *21*, 359–367. [CrossRef] [PubMed]
16. Sarkar, P.; Reichman, C.; Saleh, T.; Birge, R.B.; Kalodimos, C.G. Proline cis-trans isomerization controls autoinhibition of a signaling protein. *Mol. Cell* **2007**, *25*, 413–426. [CrossRef] [PubMed]
17. Schiene-Fischer, C.; Aumüller, T.; Fischer, G. Peptide bond cis/trans isomerases: A biocatalysis perspective of conformational dynamics in proteins. *Top. Curr. Chem.* **2013**, *328*, 35–67. [CrossRef]
18. Göthel, S.F.; Marahiel, M.A. Peptidyl-prolyl cis-trans isomerases, a superfamily of ubiquitous folding catalysts [Review]. *Cell. Mol. Life Sci.* **1999**, *55*, 423–436. [CrossRef] [PubMed]
19. Evans, P.A.; Dobson, C.M.; Kautz, R.A.; Hatfull, G.; Fox, R.O. Proline isomerism in staphylococcal nuclease characterized by NMR and site-directed mutagenesis. *Nature* **1987**, *329*, 266–268. [CrossRef] [PubMed]
20. Higgins, K.A.; Craik, D.J.; Hall, J.G.; Andrews, P.R. Cis-trans isomerization of the proline residue in insulin studied by 13C NMR spectroscopy. *Drug Des. Deliv.* **1988**, *3*, 159–170. [PubMed]
21. Chazin, W.J.; Kördel, J.; Drakenberg, T.; Thulin, E.; Brodin, P.; Grundström, T.; Forsén, S. Proline isomerism leads to multiple folded conformations of calbindin D9k: Direct evidence from two-dimensional NMR spectroscopy. *Proc. Natl. Acad. Sci. USA* **1989**, *86*, 2195–2198. [CrossRef]
22. Feng, Y.; Hood, W.F.; Forgey, R.W.; Abegg, A.L.; Caparon, M.H.; Thiele, B.R.; Leimgruber, R.M.; McWherter, C.A. Multiple conformations of a human interleukin-3 variant. *Protein Sci.* **1997**, *6*, 1777–1782. [CrossRef] [PubMed]
23. Adjadj, E.; Naudat, V.; Quiniou, E.; Wouters, D.; Sautiere, P.; Craescu, C.T. Solution structure of Lqh-8/6, a toxin-like peptide from a scorpion venom--structural heterogeneity induced by proline cis/trans isomerization. *Eur. J. Biochem./FEBS* **1997**, *246*, 218–227. [CrossRef] [PubMed]
24. Alexandrescu, A.T.; Hinck, A.P.; Markley, J.L. Coupling between local structure and global stability of a protein: Mutants of stapylococcal nuclease. *Biochemistry* **1990**, *29*, 4516–4525. [CrossRef] [PubMed]
25. Weininger, U.; Jakob, R.P.; Eckert, B.; Schweimer, K.; Schmid, F.X.; Balbach, J. A remote prolyl isomerization controls domain assembly via a hydrogen bonding network. *Proc. Natl. Acad. Sci. USA* **2009**, *106*, 12335–12340. [CrossRef] [PubMed]
26. Jakob, R.P.; Schmid, F.X. Energetic coupling between native-state prolyl isomerization and conformational protein folding. *J. Mol. Biol.* **2008**, *377*, 1560–1575. [CrossRef] [PubMed]
27. Jakob, R.P.; Schmid, F.X. Molecular determinants of a native-state prolyl isomerization. *J. Mol. Biol.* **2009**, *387*, 1017–1031. [CrossRef] [PubMed]
28. Naganathan, A.N.; Munoz, V. Insights into protein folding mechanisms from large scale analysis of mutational effects. *Proc. Natl. Acad. Sci. USA* **2010**, *107*, 8611–8616. [CrossRef]
29. de los Rios, M.A.; Muralidhara, B.K.; Wildes, D.; Sosnick, T.R.; Marqusee, S.; Wittung-Stafshede, P.; Plaxco, K.W.; Ruczinski, I. On the precision of experimentally determined protein folding rates and phi-values. *Protein Sci.* **2006**, *15*, 553–563. [CrossRef]
30. Fersht, A.R.; Sato, S. Phi-value analysis and the nature of protein-folding transition states. *Proc. Natl. Acad. Sci. USA* **2004**, *101*, 7976–7981. [CrossRef]
31. Campos, L.A. Mutational Analysis of Protein Folding Transition States: Phi Values. *Methods Mol. Biol.* **2022**, *2376*, 3–30. [CrossRef] [PubMed]
32. Vila, J.A. Protein folding rate evolution upon mutations. *Biophys. Rev.* **2023**, *15*, 661–669. [CrossRef]
33. Sosnick, T.R.; Baxa, M.C. Collapse and Protein Folding: Should We Be Surprised that Biothermodynamics Works So Well? *Annu. Rev. Biophys.* **2025**, *54*, 17–34. [CrossRef] [PubMed]
34. Schrodinger, LLC. *The PyMOL Molecular Graphics System, version 1.3r1*; Schrodinger, LLC: New York, NY, USA, 2010.
35. Holliger, P.; Riechmann, L.; Williams, R.L. Crystal structure of the two N-terminal domains of g3p from filamentous phage fd at 1.9 Angström: Evidence for conformational lability. *J. Mol. Biol.* **1999**, *288*, 649–657. [CrossRef] [PubMed]
36. Adereth, Y.; Champion, K.J.; Hsu, T.; Dammai, V. Site-directed mutagenesis using Pfu DNA polymerase and T4 DNA ligase. *Biotechniques* **2005**, *38*, 864+866+868. [CrossRef] [PubMed]
37. Santoro, M.M.; Bolen, D.W. Unfolding free energy changes determined by the linear extrapolation method. 1. Unfolding of phenylmethanesulfonyl a- chymotrypsin using different denaturants. *Biochemistry* **1988**, *27*, 8063–8068. [CrossRef] [PubMed]
38. Privalov, P.L.; Gill, S.J. Stability of protein structure and hydrophobic interaction. *Adv. Protein Chem.* **1988**, *39*, 191–234.
39. Zarrine-Afsar, A.; Davidson, A.R. The analysis of protein folding kinetic data produced in protein engineering experiments. *Methods* **2004**, *34*, 41–50. [CrossRef] [PubMed]
40. Delaglio, F.; Grzesiek, S.; Vuister, G.W.; Zhu, G.; Pfeifer, J.; Bax, A. NMRPipe: A multidimensional spectral processing system based on UNIX pipes. *J. Biomol. NMR* **1995**, *6*, 277–293. [CrossRef]

41. Johnson, B.A. Using NMRView to visualize and analyze the NMR spectra of macromolecules. *Methods Mol. Biol.* **2004**, *278*, 313–352. [CrossRef] [PubMed]
42. Jakob, R.P.; Zierer, B.K.; Weininger, U.; Hofmann, S.D.; Lorenz, S.H.; Balbach, J.; Dobbek, H.; Schmid, F.X. Elimination of a cis-proline-containing loop and turn optimization stabilizes a protein and accelerates its folding. *J. Mol. Biol.* **2010**, *399*, 331–346. [CrossRef] [PubMed]
43. Bax, A.; Clore, G.M.; Gronenborn, A.M. 1H 1H correlation via isotropic mixing of 13C magnetization, a new three-dimensional approach for assigning 1H and 13C spectra of 13C-enriched proteins. *J. Magn. Reson.* **1990**, *1969*, 7. [CrossRef]
44. Cornilescu, G.; Delaglio, F.; Bax, A. Protein backbone angle restraints from searching a database for chemical shift and sequence homology. *J. Biomol. NMR* **1999**, *13*, 289–302. [CrossRef] [PubMed]
45. Linge, J.P.; Habeck, M.; Rieping, W.; Nilges, M. ARIA: Automated NOE assignment and NMR structure calculation. *Bioinformatics* **2003**, *19*, 315–316. [CrossRef] [PubMed]
46. Fersht, A.R. From covalent transition states in chemistry to noncovalent in biology: From beta- to Phi-value analysis of protein folding. *Q. Rev. Biophys.* **2024**, *57*, e4. [CrossRef] [PubMed]
47. Martin, A.; Schmid, F.X. Evolutionary stabilization of the gene-3-protein of phage fd reveals the principles that govern the thermodynamic stability of two-domain proteins. *J. Mol. Biol.* **2003**, *328*, 863–875. [CrossRef] [PubMed]
48. Kather, I.; Jakob, R.; Dobbek, H.; Schmid, F.X. Changing the determinants of protein stability from covalent to non-covalent interactions by in vitro evolution: A structural and energetic analysis. *J. Mol. Biol.* **2008**, *381*, 1040–1054. [CrossRef]
49. Riddle, D.S.; Grantcharova, V.P.; Santiago, J.V.; Alm, E.; Ruczinski, I.; Baker, D. Experiment and theory highlight role of native state topology in SH3 folding. *Nat. Struct. Biol.* **1999**, *6*, 1016–1024. [PubMed]
50. Garcia-Mira, M.M.; Boehringer, D.; Schmid, F.X. The folding transition state of the cold shock protein is strongly polarized. *J. Mol. Biol.* **2004**, *339*, 555–569. [CrossRef] [PubMed]
51. Itzhaki, L.S.; Otzen, D.E.; Fersht, A.R. The structure of the transition state for folding of chymotrypsin inhibitor 2 analysed by protein engineering methods: Evidence for a nucleation-condensation mechanism for protein folding. *J. Mol. Biol.* **1995**, *254*, 260–288. [CrossRef] [PubMed]
52. Grantcharova, V.; Alm, E.J.; Baker, D.; Horwich, A.L. Mechanisms of protein folding. *Curr. Opin. Struct. Biol.* **2001**, *11*, 70–82. [CrossRef] [PubMed]
53. Kim, D.E.; Fisher, C.; Baker, D. A breakdown of symmetry in the folding transition state of protein L. *J. Mol. Biol.* **2000**, *298*, 971–984. [CrossRef] [PubMed]
54. Gianni, S.; Jemth, P. Conserved nucleation sites reinforce the significance of Phi value analysis in protein-folding studies. *IUBMB Life* **2014**, *66*, 449–452. [CrossRef]
55. Troilo, F.; Bonetti, D.; Camilloni, C.; Toto, A.; Longhi, S.; Brunori, M.; Gianni, S. Folding Mechanism of the SH3 Domain from Grb2. *J. Phys. Chem. B* **2018**, *122*, 11166–11173. [CrossRef]
56. Schmidpeter, P.A.M.; Rheinberger, J.; Nimigean, C.M. Prolyl isomerization controls activation kinetics of a cyclic nucleotide-gated ion channel. *Nat. Commun.* **2020**, *11*, 6401. [CrossRef] [PubMed]
57. Wang, L.; Yang, F.; Zhang, D.; Chen, Z.; Xu, R.M.; Nierhaus, K.H.; Gong, W.; Qin, Y. A conserved proline switch on the ribosome facilitates the recruitment and binding of trGTPases. *Nat. Struct. Mol. Biol.* **2012**, *19*, 403–410. [CrossRef]
58. Sarkar, P.; Saleh, T.; Tzeng, S.R.; Birge, R.B.; Kalodimos, C.G. Structural basis for regulation of the Crk signaling protein by a proline switch. *Nat. Chem. Biol.* **2011**, *7*, 51–57. [CrossRef] [PubMed]
59. Eckert, B.; Martin, A.; Balbach, J.; Schmid, F.X. Prolyl isomerization as a molecular timer in phage infection. *Nat. Struct. Mol. Biol.* **2005**, *12*, 619–623. [CrossRef]
60. Lu, K.P.; Finn, G.; Lee, T.H.; Nicholson, L.K. Prolyl cis-trans isomerization as a molecular timer. *Nat. Chem. Biol.* **2007**, *3*, 619–629. [CrossRef] [PubMed]

biomolecules

MDPI

Article

Exploring the Gating Mechanism of the Human Copper Transporter, hCtr1, Using EPR Spectroscopy

Shahaf Peleg †, Shelly Meron †, Yulia Shenberger, Lukas Hofmann, Lada Gevorkyan-Airapetov and Sharon Ruthstein *

Department of Chemistry and Institute of Nanotechnology and Advanced Materials, Faculty of Exact Sciences, Bar-Ilan University, Ramat-Gan 5290002, Israel; shahaf.peleg@biu.ac.il (S.P.); shelly.meron@biu.ac.il (S.M.); yulia.shteinbok@biu.ac.il (Y.S.); lukas.hofmman@biu.ac.il (L.H.); gavorkl@biu.ac.il (L.G.-A.)

* Correspondence: sharon.ruthstein@biu.ac.il; Tel.: +972-3-7384329

† These authors contributed equally to this work.

Abstract: Ctr1 is a membrane-spanning homotrimer that facilitates copper uptake in eukaryotic cells with high affinity. While structural details of the transmembrane domain of human Ctr1 have been elucidated using X-ray crystallography and cryo-EM, the transfer mechanisms of copper and the conformational changes that control the gating mechanism remain poorly understood. The role of the extracellular N-terminal domains is particularly unclear due to the absence of a high-resolution structure of the full-length hCtr1 protein and limited biochemical and biophysical characterization of the transporter in solution and in cell. In this study, we employed distance electron paramagnetic resonance to investigate the conformational changes of the extracellular N-terminal domain of full-length hCtr1, both in vitro and in cells, as a function of Cu(I) binding. Our results demonstrate that at specific Cu(I) concentrations, the extracellular chains move closer to the lumen to facilitate copper transfer. Additionally, while at these concentrations the intracellular part is penetrating the lumen, suggesting a ball-and-chain gating mechanism. Moreover, this phenomenon was observed for both reconstituted protein in micelles and in native cell membranes. However, the measured distance values were slightly different, suggesting that the membrane's characteristics and therefore its lipid composition also impact and even regulate the gating mechanism of hCtr1.

Keywords: copper transporter; copper metabolism; hCtr1; EPR spectroscopy

1. Introduction

The precise regulation of ion transport across biological membranes is crucial for all living cells. Ion channels and transporters are large transmembrane proteins that facilitate the selective movement of small inorganic ions. Resolving the mechanisms of the capturing, gating, and releasing of specific ions, which are governed by conformational changes within these proteins, is essential for understanding their cellular function [1–3].

The specificity of ion and ligand transporters is determined by the extracellular and intracellular domains involved in the transfer process. These domains often lack a defined secondary structure and are disordered, making it challenging to analyze the conformational changes they undergo during ligand or ion binding and transport. This dilemma impedes investigations of the gating mechanisms of these transporters. Electron paramagnetic resonance (EPR) spectroscopy has proven to be a powerful biophysical tool for obtaining high-resolution insights into these intricate biological systems. Recent work by the Cafiso and Pliotas groups has demonstrated that EPR spectroscopy can effectively track in situ conformational changes in the extracellular domains of membrane proteins within their native environment [4–6].

The human copper transporter, hCtr1, is the main gatekeeper of copper ions into the cells [7–9]. Copper is essential for cell survival; however, when its concentration is not tightly regulated, it can lead to toxicity and cell death. Therefore, the cellular systems hold restricted regulation systems, controlled by specific proteins that should shuttle the copper ions to specific subcellular locations. hCtr1 fulfills three distinct roles. In its first role, hCtr1 accumulates copper with an oxidation state of +2 from blood carrier proteins, such as human serum albumin [10–12]. In a reducing environment, the extracellular domain of hCtr1 facilitates the reduction of Cu(II) to Cu(I). In its last role, hCrt1 transports Cu(I) into the cell, where specific Cu(I) chaperones deliver it to the appropriate subcellular pathways [13–16]. The extracellular hCtr1 domain is characterized by His-rich sites [10–12], ^{1}MDHSHH and ^{22}HHH segments, and Met-based motifs, ^{7}MGMSYM and ^{41}MMMPM [17–19], that coordinate copper in an oxidation state of Cu(II) and Cu(I), respectively. Cryo-EM and X-ray crystallography have provided structural information of the hCtr1 transmembrane domain [15,20]. However, the extracellular and intracellular domains are missing from these structures. Recently, all-atom molecular dynamics (MD) simulations suggested that the extracellular domain is disordered and, upon Cu(I) binding, the extracellular domain approaches the selectivity filter, which leads to the opening of the transporter and conformational changes in the transmembrane helices [21].

We recently showed that each monomer of the hCtr1 extracellular domains binds two Cu(II) ions, resulting in a total of six Cu(II) ions per hCtr1 trimer. These results were derived from various in vitro EPR, UV-Vis measurements, and MD simulations on the full-length hCtr1 protein [22]. We also showed that a hCtr1 monomer can coordinate up to five Cu(I) ions and that the intracellular domain of hCtr1 occupies various conformational states as a function of Cu(I) concentration. More specifically, the intracellular domain is highly dynamic in its apo-state; however, at a ratio of 2–3 Cu(I):hCtr1 monomer, the intracellular C-terminal tail is folded inside the hCtr1 lumen and a homogeneous rigid structure is obtained. Subsequently, the C-terminal is released from the hCtr1 lumen upon increasing the Cu(I) concentration.

This study aims to experimentally target conformational changes in the extracellular domain of the full-length hCtr1 protein in vitro and in situ using distance EPR measurements. Paramagnetic sites were added to follow the conformational changes of the hCtr1 extracellular domain using EPR measurements. Cu(II) is a paramagnetic metal ion and in general its binding to the extracellular domain of hCtr1 can be used to gain structural information on the extracellular domain. However, since these measurements are performed in the presence of Cu(I) ions, there might be a redox reaction between them; moreover, Cu(II) is unstable in the reducing environment of the cell and is readily reduced to Cu(I) [23,24]. In order to ensure that Cu(II) bound to the extracellular domain and was not reduced, Cu(II) was protected with a nitrilamino acid (NTA) ligand (Figure 1). The group of Saxena was the first to introduce this methodology for EPR measurements [25,26] and the group of Bode has shown the high affinity of Cu(II)-NTA to His-rich sites [27–29]. The advantage of such a spin labeling methodology is that it is positioned on the backbone of the protein, with little flexibility, which leads to a four- to five-times narrower distance distribution function as compared to the common nitroxide spin labels [30]. In a recent manuscript, we illustrated how Cu(II)-NTA and dHis sites can be employed to detect conformational changes in proteins within the cellular environment [24]. Furthermore, previously we showed that Cu(II)-NTA binds to hCtr1 with a coordination site similar to that of free Cu(II) and involves at least one histidine residue from the extracellular domain of hCtr1 [3,24].

Moreover, these two recent publications demonstrated an almost complete absence of non-specific Cu(II)-NTA binding.

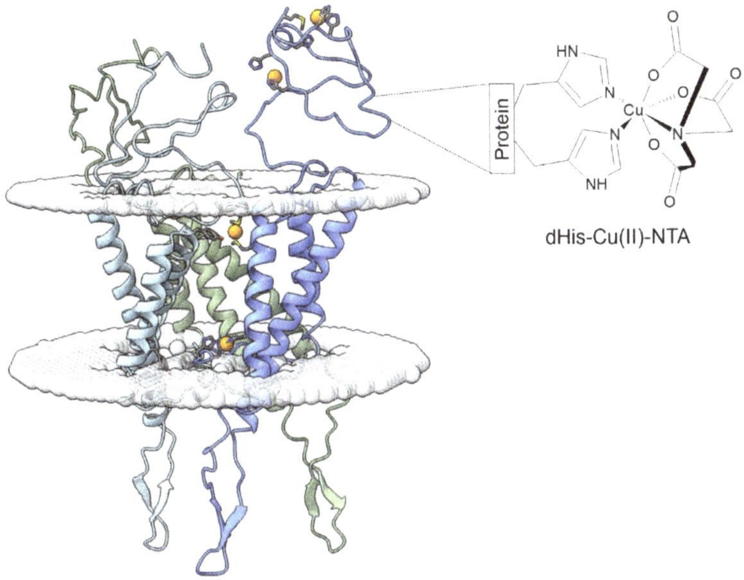

Figure 1. Schematic drawing of hCtr1 trimer. The orange balls represent Cu(I) ions. The zoom on the right-hand side illustrates the used histidine residues in complex with Cu(II)-NTA spin labeling used as spin label. The trimeric hCtr1 model is based on the structure of Ctr1 from *Salmo salar* PDB-ID: 6M98 [20]. The model was generated with Swiss-Model and prepared with Pymol (Version 2.6, Schrödinger, LLC., New York, NY, USA) and UCSF ChimeraX (Version 1.9, UCSF, CA, USA) [31].

Here, we utilized the Cu(II)-NTA spin-labeling technique coupled with distance EPR measurements to observe conformational changes in the extracellular domain of hCtr1. This was accomplished for both in vitro with protein reconstituted in micelles and in situ with overexpressed protein in *Sf9* insect cells. This methodology was employed to investigate the gating mechanism of hCtr1 in response to varying Cu(I) concentrations. The results highlighted the sensitivity of EPR spectroscopy in tracking conformational changes in disordered domains, offering valuable insights into the selectivity and transport mechanisms of ion-gated channels and transporters.

2. Materials and Methods

2.1. Cloning, Expression, and Purification of hCtr1 for In Vitro Experiments

The expression of hCtr1 was described in detail in our previous publications [3,22]. In short, wild-type (WT) hCtr1 was generated by PCR amplification and inserted into a modified pFastBac (pK503-9) vector with an N-terminal FLAG tag. To produce baculovirus for hCtr1 expression, recombinant bacmid DNA was extracted and transfected into *Sf9* insect cells (Expression Systems, LLC, Davis, CA, USA) using Cellfectin II Reagent (Thermo Fisher, Airport City, IL, USA), following the Bac-to-Bac instruction manual (Thermo Fisher, Airport City, IL, USA). The cells were cultured at 27 °C for three days and then harvested and re-suspended in a buffer containing 400 mM NaCl, 10% glycerol, and 20 mM HEPES (Sigma-Aldrich, Rehovot, IL, USA) at pH 7.4. After lysis, the resulting pellet was re-suspended in a buffer containing 1.5% Triton X-100, 200 mM NaCl, 10% glycerol, and 20 mM HEPES (pH 7.4) and incubated overnight at 4 °C. After incubation, the suspension was centrifuged again at 40,000 rpm for 40 min. The supernatant was supplemented with 3 mM $CaCl_2$ and loaded onto an anti-FLAG M1 agarose affinity gel column (Sigma-Aldrich, Rehovot, IL, USA) overnight at 4 °C, pre-equilibrated with TBS buffer (150 mM NaCl, 50 mM Tris-HCl, pH 7.4). Finally, the column was washed with TBS buffer and eluted with a buffer containing 5 mM EDTA.

To prepare the Cu(II)-NTA solution, 10 mM Cu(II) was combined with 10 mM NTA and mixed overnight. Subsequently, 240 μM Cu(II)-NTA was added to 120 μM purified hCtr1 solution and incubated overnight at 4 °C.

The addition of Cu(I) was performed as follows: concentrated Cu(I) solution (30 mM) was first prepared using Tetrakis(acetonitrile)copper(I) hexafluorophosphate (Sigma-Aldrich, Rehovot, IL, USA) dissolved in dry acetonitrile (HPLC grade) under anaerobic conditions. The Cu(I) solution was then added at different volumes to 120 μM hCtr1 monomer to give 1:1 Cu(I):hCtr1, 3:1 Cu(I):hCtr1, 5:1 Cu(I):hCtr1 ratios. Twenty percent glycerol was added to all samples.

2.2. hCtr1 In Situ and Cell Membrane Fragment Experiments

For in situ measurements according to the protocol described in [3], after the incubation of *Sf9* insect cells and hCtr1 expression, the cells were centrifuged at 1000 rpm for 5 min at 21 °C and the resulting pellet was resuspended in 30 mL of insect cell medium for a final concentration of 28.3×10^6 cells/mL. The suspension was then divided into ten test tubes, each containing 5 mL of medium with intact cells. A 250 μM Cu(II)-NTA solution was added to each tube and the samples were incubated overnight at room temperature with shaking. Following incubation, the cells were washed twice with fresh medium and subsequently divided into four samples for further analysis. Aliquots of 30 mM Cu(I) solution were then added into two samples at concentrations of 360 μM and 600 μM. A sample from each concentration was incubated for half an hour and two hours respectively. Twenty percent glycerol was added to all samples. Additional samples were lysed and subjected to ultracentrifugation at 40,000 rpm and 4 °C for 45 min.

2.3. Q-Band Double Electron–Electron Resonance (DEER) Experiments

DEER experiments ($\pi/2(\nu_{obs}) - \tau 1 - \pi(\nu_{obs}) - t' - \pi(\nu_{pump}) - (\tau 1 + \tau 2 - t') - \pi(\nu_{obs}) - \tau 2 -$ echo) were carried out with rectangular pulses without an AWG system at 20 ± 1.0 K on a Q-band Elexsys E580 spectrometer (equipped with a 2 mm probe head). A two-step phase cycle was employed on the first pulse. The echo was measured as a function of t', whereas $\tau 1$ was set to 200 ns. τ_2 was between 2500–4000 ns and was chosen based on the echo intensity and relaxation time; the exact value of τ_2 was taken into consideration in the distance distribution error analysis, as implemented in DeerAnalysis. The durations of the observer $\pi/2$ and π pulses were 14 and 28 ns, respectively, while the π pump pulse was also 28 ns. The observer frequency was constant and set to 33.85 GHz for all experiments, while the pump frequency was set to 33.74 GHz and the magnetic field was 11,680 G. Previous studies have shown that within the range of 500 G, orientational effects can be ignored at the $g_{\perp}$ region [32,33]. The data were analyzed using the DeerAnalysis 2019 program (ETH, Switzerland). Tikhonov regularization and DeerNet were also used to analyze the data [34]. The x-axis in the distance distribution was corrected based on the g-value and was multiplied by $(g_{nitroxide}/g_{\perp,Cu})^{0.67} = 0.986$; where $g_{nitroxide} = 2.0057$, and $g_{\perp,Cu} = 2.05$ [32].

3. Results

The full-length hCtr1 was expressed in insect cells and Cu(II)-NTA was added either to cells (expressing hCtr1) or to purified protein at a ratio of 2:1 Cu(II)-NTA:hCtr1, as was also described in our previous publication [3]. Cu(I) was then added to the cells or hCtr1 monomer at different ratios as described in the materials and methods section under anaerobic conditions. An SDS-Gel picture of the purified hCtr1 is shown in Figure S1, Supplementary Materials. Western blot experiments were carried out to ensure that the presence of Cu(II)-NTA and Cu(I) does not affect hCtr1 expression (Figure S2, Supplementary Materials).

Double electron–electron resonance (DEER) pulsed EPR distance measurements were conducted to evaluate the distance distributions between Cu(II)-NTA sites of the hCtr1 extracellular domains. Initially, the DEER measurements were performed on purified hCtr1 reconstituted in triton micelles (Figure 2) as a function of [Cu(I)]. The DEER data suggest for the apo hCtr1 (no Cu(I) added) a bimodal distribution between 2.0–3.5 nm and an additional distribution around a longer distance of 5.6 nm. The addition of small amounts of Cu(I) in a ratio of 1:1 Cu(I):hCtr1 monomer results in distance distributions around 2.1 nm and between 4.0–5.0 nm. However, the addition of larger amounts of Cu(I) to the solution in a ratio of 3Cu(I):hCtr1 monomer results in a single distance distribution function around 1.8 nm (repeated experiments for this concentration are provided in Figure S3, Supplementary Materials). The addition of 5Cu(I):hCtr1 monomer reveals only the distribution at longer distances between 4.0–5.0 nm. This phenomenon is very similar to what was detected for the intracellular domain of hCtr1 at a 3Cu(I):hCtr1 ratio. While at other copper concentrations lower or higher than 3Cu(I):hCtr1 monomer, various distance distribution functions appear at longer distances [22]. In this study, MD simulations suggested that such homogeneous small-distance distribution can only occur when the three intracellular tails of hCtr1 enter the lumen to transfer Cu(I) from the lumen to the cell [22]. Recently, MD simulations on the extracellular domain of hCtr1 also suggested that the extracellular domain must approach the hCtr1 funnel to allow for Cu(I) ion transfer [21]. Altogether, the EPR and MD data support the notion that at a ratio of 3Cu(I):hCtr1, both the extracellular and the intracellular domains should approach the hCtr1 lumen to allow fast copper transfer into the cell.

Next, for in situ measurements, Q-band DEER experiments were performed with Cu(II)-NTA bound to hCtr1 in the cells. To begin, 240 μM Cu(II)-NTA was added to the cells (in our expression protocol, the amount of purified hCtr1 was in the range between 120–140 μM) and Cu(I) was added to the cells at different concentrations (360/600 μM, corresponding to 3Cu(I):hCtr1 monomer or 5Cu(I):hCtr1 monomer). The DEER data for the cells are presented in Figure 3. In the absence of Cu(I), the DEER data suggest distributions of around 4.8–6.0 nm. The smaller distances that were observed for the purified protein in micelles between 2.0–3.5 nm were not detected in the cells. As previously discussed [3], we conclude that in the native environment, the extracellular chains are in proximity to the functional group of the phospholipids and associated with the membrane, resulting in a more distant organization of the three disordered N-termini. Conversely, once the protein is reconstituted in micelles, this anchoring is disrupted, which led to the observed close configuration of the three extracellular domains of hCtr1.

DEER measurements were next performed in the presence of Cu(I). The addition of copper at a ratio of 3Cu(I):hCtr1 suggested a distribution of around 1.8 nm, similar to the purified protein at this concentration (repeated experiments for this concentration are provided in Figure S4, Supplementary Materials), while at higher amounts of copper, this distribution disappears and distributions between 3.8–5.5 nm appear. Although there are some differences in the precise values of the distances between the in vitro and in situ EPR measurements, the observation that the extracellular chains are approaching each other at the same specific concentration in both systems is striking. The changes in the distance values highlight the importance of the lipids or surfactants in the function of gating transporters [35,36].

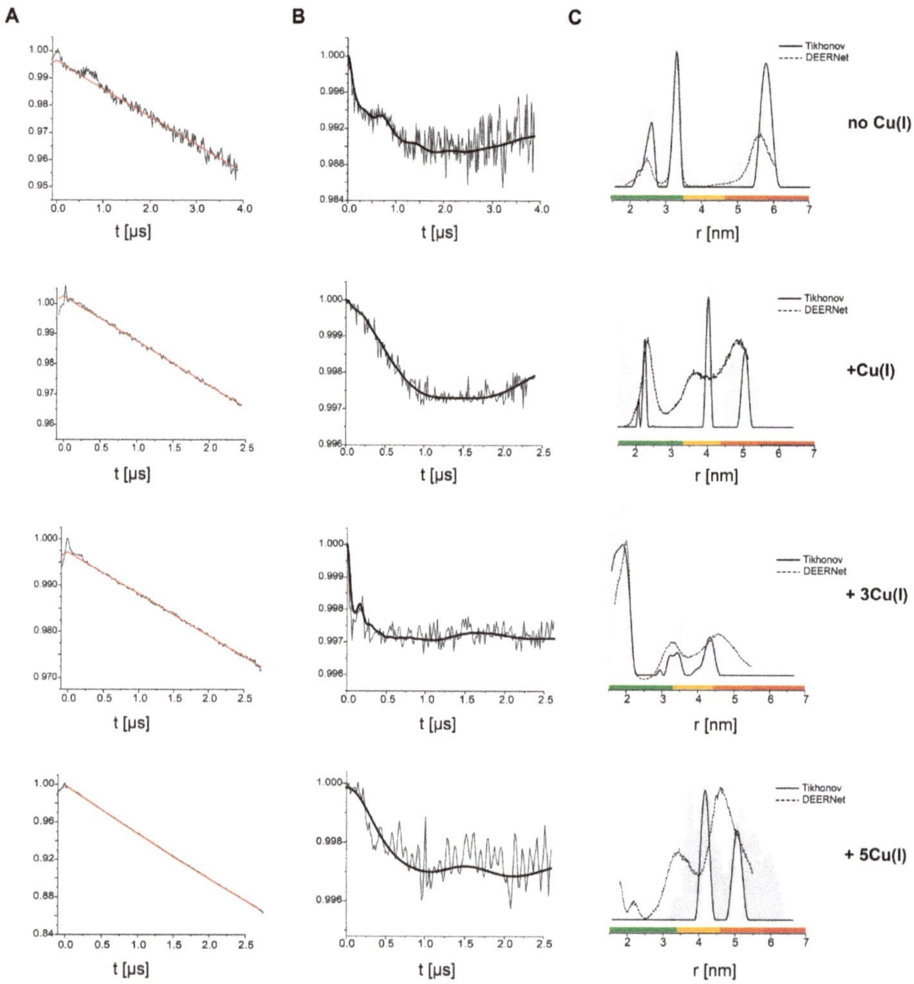

Figure 2. Q-band DEER measurements on purified hCtr1 upon Cu(I) coordination. (**A**) Raw time domain DEER data (black) and the background function (red). (**B**) DEER time domain data after background correction and the corresponding fit. (**C**) The corresponding distance distributions. The data were analyzed using the DeerAnalysis program using Tikhonov regularization with a regularization parameter of 20 (solid black lines) and DEERNet (dashed black lines). Distance distribution validation considered white noise, background start. and dimensionality. The differences between the Tikhonov analysis and DEERNet are within the margin of error. The color bar indicates reliability ranges (green: shape reliable; yellow: mean and width reliable; orange: mean reliable; red: no quantification possible). The data were acquired at 20 K, where 2:1 Cu(II)-NTA to 120 μM hCtr1 monomer in HEPES buffer, pH 7.4, and 20% glycerol were added. Cu(I) was added at different ratios as compared to hCtr1 monomer. The data for the apo hCtr1 (no Cu(I)) were reproduced from Meron et al. 2024 with permission from ACS [3].

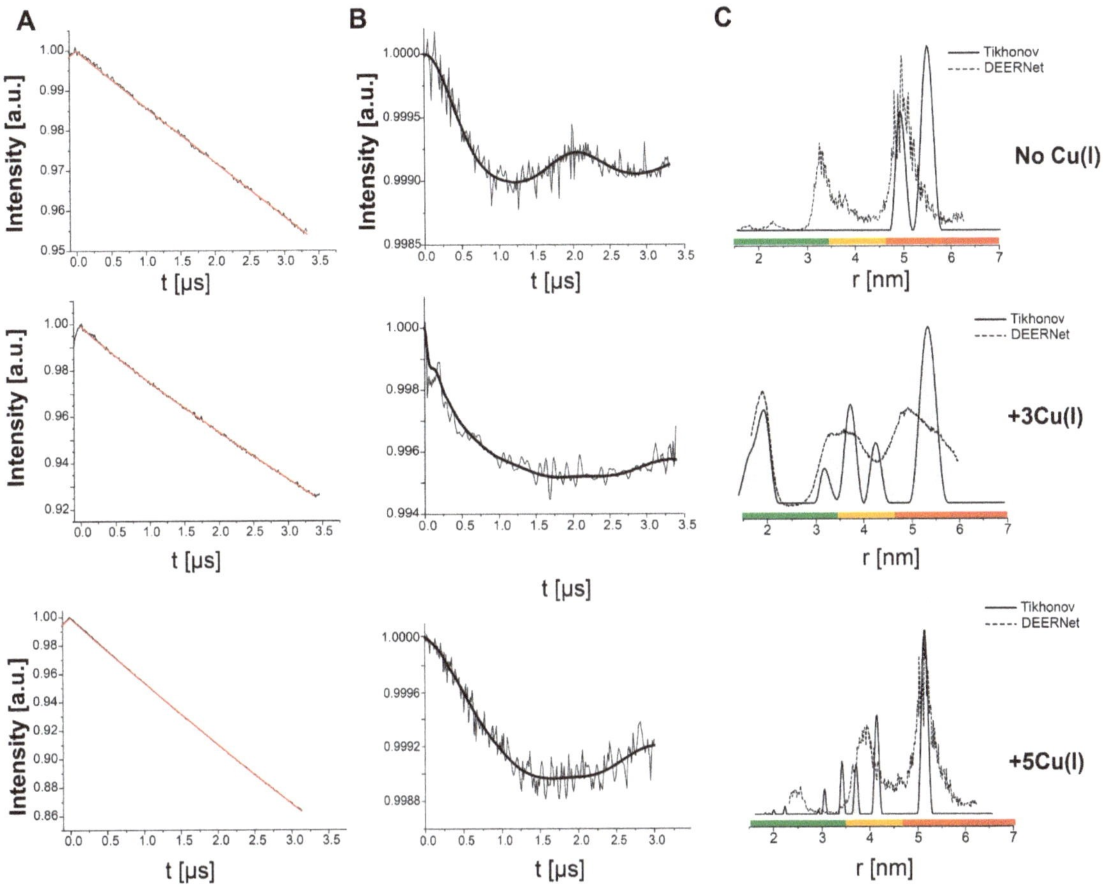

Figure 3. Q-band DEER measurements on Cu(II)-NTA in situ of hCtr1 upon Cu(I) coordination. (**A**) Raw time domain DEER data (black) and the background function (red). (**B**) DEER time domain data after background correction and the corresponding fit. (**C**) The corresponding distance distributions. The data were analyzed using the DeerAnalysis program using Tikhonov regularization with a regularization parameter of 20 (solid black lines) and DEERNet (dashed black lines). Distance distribution validation considered white noise, background start, and dimensionality. The differences between the Tikhonov analysis and DEERNet are within the margin of error. The color bar indicates reliability ranges (green: shape reliable; yellow: mean and width reliable; orange: mean reliable; red: no quantification possible). The data were acquired at 20 K, where 240 μM Cu(II)-NTA was added to the cells, 20% glycerol, and different Cu(I) concentrations (360 μM/600 μM). The data for the apo hCtr1 (no Cu(I)) were reproduced from Meron et al. 2024 with permission from ACS [3].

In order to further verify this, we lysed the cells after adding Cu(II)-NTA and Cu(I) and EPR measurements were conducted (Figure 4). Without Cu(I), the DEER data suggest a bimodal distance distribution between 4.0–5.5 nm, similar to the in situ cells experiment. The addition of 3Cu(I):hCtr1 resulted in additional distributions with smaller distances between 2.0–4.0 nm. Adding 5Cu(I):hCtr1 led to a reduction in the contributions of the distributions around 2.0 nm, which agrees well with the findings on the purified and in situ hCtr1 experiments. Pulsed EPR measurements were also conducted on *sf9* cells without hCtr1 overexpression (Figure S5, Supplementary Materials). In this case, the echo intensity was five times lower, as the Cu(II)-NTA was washed away, confirming that there is no specific binding without hCtr1; moreover, no signal at all was detected in the cell membrane fragments. The relaxation time of the non-bound Cu(II)-NTA in the cells is much faster and its contribution to the DEER signal is negligible (Figure S5, Supplementary Materials).

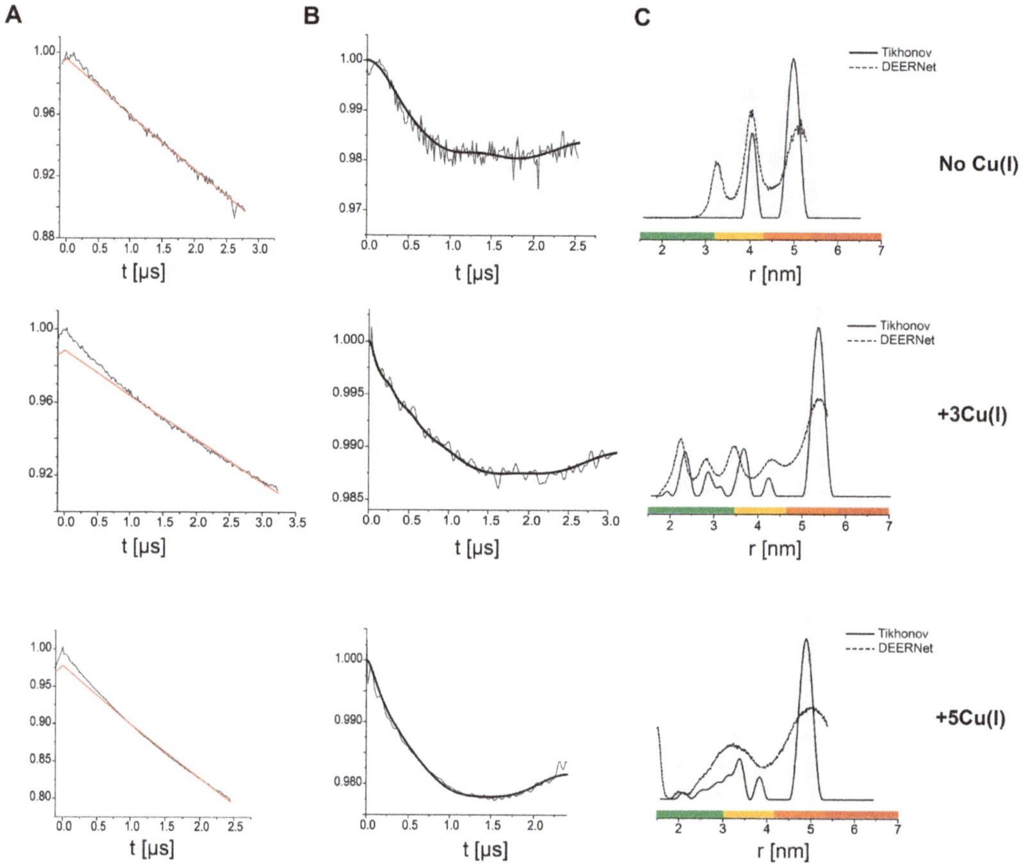

Figure 4. Q-band DEER measurements on Cu(II)-NTA bound to hCtr1 in cell membrane fragments as a function of Cu(I) coordination. (**A**) Raw time domain DEER data (black) and the background function (red). (**B**) DEER time domain data after background correction and the corresponding fit. (**C**) The corresponding distance distributions. The data were analyzed using the DeerAnalysis program using Tikhonov regularization with a regularization parameter of 20 (solid black lines) and DEERNet (dashed black lines). Distance distribution validation considered white noise, background start, and dimensionality. The differences between the Tikhonov analysis and DEERNet are within the margin of error. The color bar indicates reliability ranges (green: shape reliable; yellow: mean and width reliable; orange: mean reliable; red: no quantification possible). The data were acquired at 20 K, where 240 μM Cu(II)-NTA, 20% glycerol, and different Cu(I) concentrations (360 μM/600 μM) were added to the cells and the cells were then lysed and centrifuged. The data for the apo hCtr1 (no Cu(I)) were reproduced from Meron et al. 2024 with permission from ACS [3].

The modulation depth detected here, which is related to the excitation profile of the paramagnetic centers, is very low. In general, Cu(II) distance measurements are characterized by a low modulation depth, owing to the large spectral width of Cu(II) [37]. Here, the limitation is even larger owing to the fact that the studied system is a human membrane transporter with a comparable low expression yield [3]. Despite this limitation, the use of this spin-labeling methodology allows for comparable highly resolved DEER data, due to the orthogonal labeling, which reduces the flexibility of the spin labeling and allows narrow distance distribution functions [24,33,37]. In the purified protein, the modulation depth is ~1–2%, which is larger than that in the cellular environment (~0.1%). This suggests that less Cu(II)-NTA molecules are bound to the extracellular domain of the hCtr1 trimer in the cellular environment compared to the reconstituted hCtr1 in micelles.

Despite the experimental limitations, DEER experiments were able to follow conformational changes in vitro and in situ of the disordered domains of extracellular hCtr1 upon Cu(I) binding. Figure 5 illustrates the proposed copper gating mechanism of copper through the hCtr1 transporter. In the apo-state, the extracellular N-termini chains are further apart and the intracellular chains are in the cytoplasmic domain (opened state).

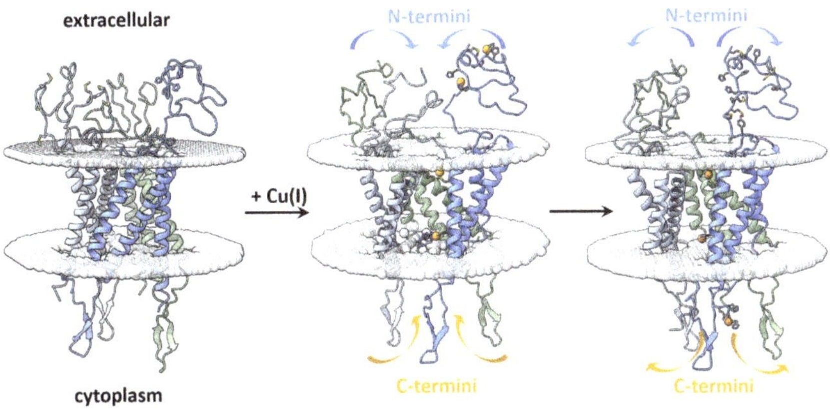

Figure 5. The copper transfer mechanism through hCtr1 transporter. Upon Cu(I) uptake, the extracellular chains move closer to each other and the intracellular chains move into the lumen (closed state). Once Cu(I) is transferred into the lumen, the extracellular chains are spread apart, with the intracellular chains pointing into the cytoplasmic domain for the transfer of copper to the various metallochaperones (opened state). This figure was created as described in Figure 1 and prepared using UCSF ChimeraX.

At specific Cu(I) concentrations, the extracellular chains move closer to each other and the intracellular chains move toward the protein lumen (closed state). At higher copper concentrations, the intracellular chains are released back to the cytoplasmic side and the extracellular domains move further apart, enabling them to scavenge more copper ions.

4. Discussion

Gaining structural information on ion transporters within the cell is a complex task, particularly when focusing on the extracellular domains that regulate specificity and gating mechanisms. These regions often lack a well-defined secondary structure and are intrinsically disordered, making it challenging to study their conformational changes. The difficulty is increased when the experiments are performed in the cellular environment and are compared to the micellar environment. In micelles, proteins may adopt conformations or interact differently compared to their behavior in the cell, potentially leading to discrepancies in the observed gating mechanisms. In contrast, studying transporters within the cell provides a more physiologically relevant setting but introduces additional variables and complexities, such as interactions with other cellular components and the dynamic nature of the cellular environment [35,36,38]. Thus, accurately capturing the structural dynamics of extracellular domains and their role in specificity and gating mechanisms requires careful consideration of these experimental contexts and the limitations they impose.

EPR spectroscopy is a powerful biophysical tool for studying conformational changes in complex biological systems and transmembrane systems that are difficult to monitor with other methods. EPR is particularly useful for tracking conformational changes in disordered domains of transporters and channels [4–6,22] or following a transcription mechanism [32,39]. We recently demonstrated the use of Cu(II)-NTA and dHis sites as spin-labeling methodologies for EPR distance measurements to investigate conformational changes of proteins in cellular conditions [3,24]. Here, we applied this approach to monitor structural changes in the extracellular domain of the human copper transporter, hCtr1. The Cu(II)-NTA spin-labeling method provides several advantages for studying hCtr1 in its cellular environment: first, multiple native histidine residues in the extracellular domains of hCtr1 serve as binding sites. Second, EPR data indicate that Cu(II)-NTA binds to hCtr1 similarly as to free Cu(II). Because NTA protects Cu(II) from reduction and from reacting with free Cu(I) ions, this spin-labeling approach allows us to track conformational changes in hCtr1 in response to Cu(I) binding. Our EPR distance measurements indicate that the extracellular chains of hCtr1 come closer together upon Cu(I) binding, both in purified hCtr1 within micelles and in the native cell membrane. Notably, at a specific ratio of three Cu(I) ions to one hCtr1 monomer, the extracellular chains exhibit a single distance distribution function. This finding suggests that all chains simultaneously move closer to the hCtr1 lumen. Previously, similar findings proposed that at this specific Cu(I) concentration, the intracellular C-terminal chains also approach the hCtr1 lumen to facilitate copper ion transfer [22].

The gating mechanisms of ion-gated transporters such as potassium and calcium channels have been extensively studied using various biophysical and computational methods. These studies indicate that gating is regulated by a synergy between the high affinity of the extracellular domain for specific ions, dynamic and conformational changes in the extracellular domain that lead to interactions with the membrane, and alterations in the orientation of transmembrane helices and the penetration of the intracellular loop into the lumen, a mechanism often referred to as the "ball-and-chain" mechanism [40,41]. This mechanism is analogous to our observations of copper gating by hCtr1. We observed transitions between open and closed states in both the extracellular and intracellular regions at specific copper concentrations. In the closed state, characterized by close interactions among all extracellular domains and the penetration of the intracellular domain into the lumen, we also noted that the membrane environment influenced these interactions. Specifically, EPR measurements in Triton micelles revealed closer interactions among the extracellular chains compared to those observed in the native cell membrane.

5. Conclusions

EPR measurements conducted on the full-length human copper transporter, hCtr1, in reconstituted protein in micelles and expressed protein in insect cells as a function of Cu(I) ion concentration suggested a gating mechanism involving opened and closed states, where in the closed state the extracellular chains approach each other and the intracellular chain is penetrating the lumen. These two states were present for the reconstituted protein in triton micelles and in the native membrane. However, the distance distribution values were slightly different between these two states, also highlighting the importance of the membrane characteristic to the gating mechanism.

Supplementary Materials: The following supporting information can be downloaded at: https://www.mdpi.com/article/10.3390/biom15010127/s1, Figure S1: SDS-PAGE analysis of purified WT-hCtr1 following silver staining. Lanes 2–6 display the elution fractions obtained from an anti-FLAG M1 agarose affinity gel column using a solution containing 5 mM EDTA and 100 μg/mL FLAG peptide; Figure S2: Western blot of WT-hCtr1 with Cu(II)-NTA complex and Cu(I)-tetrakis at different ratios; Figure S3: Various DEER time domain signals and corresponding distance distributions functions were acquired on different samples of purified hCtr1 in the presence of Cu(II)-NTA and 3Cu(I); Figure S4: Various DEER time domain signals and corresponding distance distributions functions were acquired on different samples of overexpressed hCtr1 in the cells, in the presence of Cu(II)-NTA and 3Cu(I); Figure S5: (A). Two-pulse echo decay for Cu(II)-NTA in sf9 cells with (red) and without hCtr1 expression (black). (B). DEER time domain signal for Cu(II)-NTA in sf9 cells with (red) and without hCtr1 expression (black); Figure S6. Original Western Blot image ctr1 in insect cells.

Author Contributions: Conceptualization, S.P. and S.R.; methodology, S.P., S.M., Y.S. and L.G.-A.; software, L.H.; formal analysis, S.P., S.M. and S.R.; investigation, S.P., S.M., L.H., Y.S. and L.G.-A.; resources, S.R.; writing—original draft preparation, S.P., S.M. and L.H.; writing—review and editing, S.P., L.H. and S.R. supervision, L.H. and S.R. All authors have read and agreed to the published version of the manuscript.

Funding: This research was funded by the Israel Science Foundation, ISF 212/22.

Institutional Review Board Statement: Not applicable.

Informed Consent Statement: Not applicable.

Data Availability Statement: The original contributions presented in this study are included in the article/Supplementary Materials. Further inquiries can be directed to the corresponding author.

Conflicts of Interest: The authors declare that there are no conflicts of interest.

References

1. Davis, J.T.; Okunola, O.; Quesada, R. Recent advances in the transmembrane transport of anions. *Chem. Soc. Rev.* **2010**, *39*, 3843–3862. [CrossRef] [PubMed]
2. Hoshi, T.; Zagotta, W.N.; Aldrich, R.W. Biophysical and Molecular Mechanisms of Shaker Potassium Channel Inactivation. *Science* **1990**, *250*, 533–538. [CrossRef]
3. Meron, S.; Peleg, S.; Shenberger, Y.; Hofmann, L.; Gevorkyan-Airapetov, L.; Ruthstein, S. Tracking Disordered Extracellular Domains of Membrane Proteins in the Cell with Cu(II)-Based Spin Labels. *J. Phys. Chem. B* **2024**, *128*, 8908–8914. [CrossRef]
4. Nyenhuis, D.A.; Nilaweera, T.D.; Cafiso, D.S. Native Cell Environment Constrains Loop Structure in the Escherichia coli Cobalamin Transporter BtuB. *Biophys. J.* **2020**, *119*, 1550–1557. [CrossRef]
5. Haysom, S.F.; Machin, J.; Whitehouse, J.M.; Horne, J.E.; Fenn, K.; Ma, Y.; El Mkami, H.; Bohringer, N.; Schaberle, T.F.; Ranson, N.A.; et al. Darobactin B Stabilises a Lateral-Closed Conformation of the BAM Complex in E. coli Cells. *Angew. Chem. Int. Ed. Engl.* **2023**, *62*, e202218783. [CrossRef]
6. Nyenhuis, D.A.; Nilaweera, T.D.; Niblo, J.K.; Nguyen, N.Q.; DuBay, K.H.; Cafiso, D.S. Evidence for the Supramolecular Organization of a Bacterial Outer-Membrane Protein from In Vivo Pulse Electron Paramagnetic Resonance Spectroscopy. *J. Am. Chem. Soc.* **2020**, *142*, 10715–10722. [CrossRef] [PubMed]
7. De Feo, C.J.; Aller, S.G.; Siluvai, G.S.; Blackburn, N.J.; Unger, V.M. Three-dimensional structure of the human copper transporter hCTR1. *Proc. Nat. Acad. Sci. USA* **2009**, *106*, 4237–4242. [CrossRef]
8. Zimnicka, A.M.; Maryon, E.B.; Kaplan, J.H. Human copper transporter hCTR1 mediates basolateral uptake of copper into enterocytes: Implications for copper homeostasis. *J. Biol. Chem.* **2007**, *282*, 26471–26480. [CrossRef]
9. Zhou, B.; Gitschier, J. hCTR1: A human gene for copper uptake identified by complementation in yeast. *Proc. Natl. Acad. Sci. USA* **1997**, *94*, 7481–7486. [CrossRef] [PubMed]
10. Stefaniak, E.; Plonka, D.; Drew, S.C.; Bossak-Ahmad, K.; Haas, K.L.; Pushie, M.J.; Faller, P.; Wezynfeld, N.E.; Bal, W. The N-terminal 14-mer model peptide of human Ctr1 can collect Cu(ii) from albumin. Implications for copper uptake by Ctr1. *Metallomics* **2018**, *10*, 1723–1727. [CrossRef] [PubMed]
11. Bossak, K.; Drew, S.C.; Stefaniak, E.; Plonka, D.; Bonna, A.; Bal, W. The Cu(II) affinity of the N-terminus of human copper transporter CTR1: Comparison of human and mouse sequences. *J. Inorg. Biochem.* **2018**, *182*, 230–237. [CrossRef] [PubMed]
12. Shenberger, Y.; Shimshi, A.; Ruthstein, S. EPR spectroscopy shows that the blood carrier protein, human serum albumin, closely interacts with the N-terminal domain of the copper transporter, Ctr1. *J. Phys. Chem. B* **2015**, *119*, 4824–4830. [CrossRef]
13. Maryon, E.B.; Molloy, S.A.; Ivy, K.; Yu, H.; Kaplan, J.H. Rate and regulation of copper transport by human copper transporter (hCTR1). *J. Biol. Chem.* **2013**, *288*, 18035–18046. [CrossRef]
14. Du, X.; Li, H.; Wang, X.; Liu, Q.; Ni, J.; Sun, H. Kinetics and thermodynamics of metal binding to the N-terminus of a human copper transporter, hCTR1. *Chem. Commun.* **2013**, *49*, 9134–9136. [CrossRef]

15. Aller, S.G.; Unger, V.M. Projection structure of the human copper transporter CTR1 at 6A resolution structure reveals a compact trimer with a novel channel-like architecture. *Proc. Nat. Acad. Sci. USA* **2006**, *103*, 3627–3632. [CrossRef]
16. Lee, J.; Pena, M.M.O.; Nose, Y.; Thiele, D.J. Biochemical Characterization of the Human Copper Transporter Ctr1. *J. Biol. Chem.* **2002**, *277*, 4380–4387. [CrossRef]
17. Jiang, J.; Nadas, I.A.; Kim, J.M.; Franz, K.J. A mets motif peptide found in copper transport proteins selectively binds Cu(I) with methionine-only coordination. *Inorg. Chem.* **2005**, *44*, 9787–9794. [CrossRef]
18. Shenberger, Y.; Marciano, O.; Gottlieb, H.; Ruthstein, S. Insights into the N-terminal Cu(II) and Cu(I) binding sites of the human copper transporter CTR1. *J. Coord. Chem.* **2018**, *71*, 1985–2002. [CrossRef]
19. Magistrato, A.; Pavlin, M.; Qasem, Z.; Ruthstein, S. Copper trafficking in eukaryotic systems: Current knowledge from experimental and computational efforts. *Curr. Opin. Struct. Biol.* **2019**, *58*, 26–33. [CrossRef] [PubMed]
20. Ren, F.; Logeman, B.L.; Zhang, X.; Liu, Y.; Thiele, D.J.; Yuan, P. X-ray structures of the high-affinity copper transporter Ctr1. *Nat. Commun.* **2019**, *10*, 1386. [CrossRef] [PubMed]
21. Aupic, J.; Lapenta, F.; Janos, P.; Magistrato, A. Intrinsically disordered ectodomain modulates ion permeation through a metal transporter. *Proc. Natl. Acad. Sci. USA* **2022**, *119*, e2214602119. [CrossRef] [PubMed]
22. Walke, G.; Aupic, J.; Kashoua, H.; Janos, P.; Meron, S.; Shenberger, Y.; Qasem, Z.; Gevorkyan-Airapetov, L.; Magistrato, A.; Ruthstein, S. Dynamical interplay between the human high-affinity copper transporter hCtr1 and its cognate metal ion. *Biophys. J.* **2022**, *121*, 1194–1204. [CrossRef] [PubMed]
23. Tsang, T.; Davis, C.I.; Brady, D.C. Copper biology. *Curr. Biol.* **2021**, *31*, R421–R427. [CrossRef]
24. Shenberger, Y.; Gevorkyan-Airapetov, L.; Hirsch, M.; Hofmann, L.; Ruthstein, S. An in-cell spin-labelling methodology provides structural information on cytoplasmic proteins in bacteria. *Chem. Commun.* **2023**, *59*, 10524–10527. [CrossRef] [PubMed]
25. Cunningham, T.F.; Putterman, M.R.; Desai, A.; Horne, W.S.; Saxena, S. The Double Histidine Cu^{2+}-Binding Motif: A Highly Rigid, Site-Specific Spin Probe for Electron Spin Resonance Distance Measurements. *Angew. Chem. (Int. Ed.)* **2015**, *54*, 6330–6334. [CrossRef] [PubMed]
26. Casto, J.; Bogetti, X.; Hunter, H.R.; Hasanbasri, Z.; Saxena, S. "Store-bought is fine": Sensitivity considerations using shaped pulses for DEER measurements on Cu(II) labels. *J. Magn. Reson.* **2023**, *349*, 107413. [CrossRef] [PubMed]
27. Ackermann, K.; Heubach, C.A.; Schiemann, O.; Bode, B.E. Pulse Dipolar Electron Paramagnetic Resonance Spectroscopy Distance Measurements at Low Nanomolar Concentrations: The Cu(II)-Trityl Case. *J. Phys. Chem. Lett.* **2024**, *15*, 1455–1461. [CrossRef]
28. Ackermann, K.; Wort, J.L.; Bode, B.E. Nanomolar Pulse Dipolar EPR Spectroscopy in Proteins: Cu(II)-Cu(II) and Nitroxide-Nitroxide Cases. *J. Phys. Chem. B* **2021**, *125*, 5358–5364. [CrossRef]
29. Wort, J.L.; Arya, S.; Ackermann, K.; Stewart, A.J.; Bode, B.E. Pulse Dipolar EPR Reveals Double-Histidine Motif Cu(II)-NTA Spin-Labeling Robustness against Competitor Ions. *J. Phys. Chem. Lett.* **2021**, *12*, 2815–2819. [CrossRef]
30. Gamble Jarvi, A.; Bogetti, X.; Singewald, K.; Ghosh, S.; Saxena, S. Going the dHis-tance: Site-Directed Cu(2+) Labeling of Proteins and Nucleic Acids. *Acc. Chem. Res.* **2021**, *54*, 1481–1491. [CrossRef]
31. Waterhouse, A.; Bertoni, M.; Bienert, S.; Studer, G.; Tauriello, G.; Gumienny, R.; Heer, F.T.; de Beer, T.A.P.; Rempfer, C.; Bordoli, L.; et al. SWISS-MODEL: Homology modelling of protein structures and complexes. *Nucleic Acids Res.* **2018**, *46*, W296–W303. [CrossRef]
32. Sameach, H.; Ghosh, S.; Gevorkyan-Airapetov, L.; Saxena, S.; Ruthstein, S. EPR Spectroscopy Detects Various Active State Conformations of the Transcriptional Regulator CueR. *Angew. Chem. Int. Ed. Engl.* **2019**, *58*, 3053–3056. [CrossRef]
33. Jarvi, A.G.; Ranguelova, K.; Ghosh, S.; Weber, R.T.; Saxena, S. On the Use of Q-Band Double Electron-Electron Resonance to Resolve the Relative Orientations of Two Double Histidine-Bound Cu2+ Ions in a Protein. *J. Phys. Chem. B* **2018**, *122*, 10669–10677. [CrossRef] [PubMed]
34. Worswick, S.G.; Spencer, J.A.; Jeschke, G.; Kuprov, I. Deep neural network processing of DEER data. *Sci. Adv.* **2018**, *4*, eaat5218. [CrossRef] [PubMed]
35. Gamper, N.; Shapiro, M.S. Regulation of ion transport proteins by membrane phosphoinositides. *Nat. Rev. Neurosci.* **2007**, *8*, 921–934. [CrossRef] [PubMed]
36. Levental, I.; Lyman, E. Regulation of membrane protein structure and function by their lipid nano-environment. *Nat. Rev. Mol. Cell Biol.* **2023**, *24*, 107–122. [CrossRef]
37. Hunter, H.R.; Kankati, S.; Hasanbasri, Z.; Saxena, S.K. Endogenous Cu(II) Labeling for Distance Measurements on Proteins by EPR. *Chemistry* **2024**, *30*, e202403160. [CrossRef]
38. Gu, R.X.; de Groot, B.L. Lipid-protein interactions modulate the conformational equilibrium of a potassium channel. *Nat. Commun.* **2020**, *11*, 2162. [CrossRef] [PubMed]
39. Hofmann, L.; Mandato, A.; Saxena, S.; Ruthstein, S. The use of EPR spectroscopy to study transcription mechanisms. *Biophys. Rev.* **2022**, *14*, 1141–1159. [CrossRef]

40. Sukomon, N.; Fan, C.; Nimigean, C.M. Ball-and-Chain Inactivation in Potassium Channels. *Annu. Rev. Biophys.* **2023**, *52*, 91–111. [CrossRef]
41. Fan, C.; Sukomon, N.; Flood, E.; Rheinberger, J.; Allen, T.W.; Nimigean, C.M. Ball-and-chain inactivation in a calcium-gated potassium channel. *Nature* **2020**, *580*, 288–293. [CrossRef] [PubMed]

biomolecules

MDPI

Review

Standard Sample Preparation for Serial Femtosecond Crystallography

Christina Schmidt, Kristina Lorenzen, Joachim Schulz and Huijong Han *

European XFEL GmbH, Holzkoppel 4, 22869 Schenefeld, Germany; christina.schmidt@xfel.eu (C.S.); joachim.schulz@xfel.eu (J.S.)
* Correspondence: huijong.han@xfel.eu; Tel.: +49-40-8998-6795

Abstract

The development of serial crystallography (SX), including serial synchrotron crystallography (SSX) at synchrotron sources and serial femtosecond crystallography (SFX) at X-ray free-electron lasers (XFELs), has facilitated the collection of high-resolution diffraction data from micron-sized crystals, providing unique insights into the structures and dynamics of biomolecules at room temperature. Standard samples are essential for the commissioning of new XFEL instruments and the validation of experimental setups. In this review, we summarize currently used standard proteins and describe representative microcrystal preparation workflows for four widely adopted models, lysozyme, myoglobin, iq-mEmerald, and photoactive yellow protein (PYP), drawing on established methodologies and accumulated experience from their applications at the European XFEL. By consolidating existing knowledge and integrating protocols that have been systematically refined and optimized through our experimental efforts, this review aims to provide practical guidance for the serial crystallography community, thereby enhancing reproducibility and ensuring consistent experimental performance across facilities.

Keywords: serial femtosecond crystallography; microcrystals; standard samples; protocol

1. Introduction

Serial crystallography (SX) is a revolutionary technique in structural biology that has existed for more than a decade, providing insights into the structures and dynamics of biomolecules at room temperature [1,2]. Using intense ultra-short X-ray pulses, SX made the collection of diffraction data from micron-sized crystals possible, avoiding radiation damage and enabling the capture of transient states and intermediates in biological reactions.

The success of SX experiments relies heavily on the quality of the sample [3–6]. Hence, standard samples in SX research provide critical roles. First, the commissioning of new beamline devices needs well-characterized samples. By providing consistent and reproducible diffraction patterns, standard samples help in calibrating new devices and verifying their performance. Secondly, they are required for the validation of experimental setups, ensuring that all components, from sample delivery systems to data acquisition software, function correctly.

To balance availability with practical preparation requirements, all standard samples must have robust crystallization properties. Well-documented structural data must be available for the comparability of verification measurements. Several proteins have emerged as common standards in SX research due to their ability to fulfill these criteria. These proteins not only crystallize reliably but also tolerate various sample delivery methods, data collection conditions, and experimental designs. An overview listing the most commonly used standard samples in SX experiments is given in Table 1.

The following describes several standard samples widely used in SX. Each one has specific characteristics that make it particularly suitable for instrument commissioning, method validation, or the development of time-resolved experiments.

1.1. *Lysozyme*

Lysozyme has long been the reference protein in crystallography due to its reliable crystallization behavior and well-characterized structure. In serial crystallography, its practical advantages are particularly clear because lysozyme readily forms microcrystals under a wide range of conditions, and these crystals consistently yield high-quality diffraction, making them ideal for systematic testing and method development [7–10].

The compatibility of lysozyme microcrystals with various sample delivery methods, including liquid jets [1,11], high-viscosity extrusion (HVE) [8], and fixed targets [12], has made lysozyme the protein of choice for optimizing experimental configurations. Lysozyme's reproducibility across crystallization and data collection conditions also makes it ideal for evaluating critical aspects such as radiation damage, sample consumption, and hit rates.

Altogether, its ease of crystallization, structural consistency, and experimental flexibility have made it a foundational standard in the development and refinement of serial crystallography techniques.

1.2. *Thermolysin*

Thermolysin, a thermostable metalloprotease from *Geobacillus stearothermophilus* (34.6 kDa), is widely used as a standard sample in serial crystallography, especially for establishing different sample delivery methods such as a concentric flow electrokinetic injector, nanoflow electrospin liquid jets, on-chip crystallization, acoustic injectors and novel injection matrices, and ligand-soaking experiments, amongst others [13–18].

Available as a stable, lyophilized powder, thermolysin forms microcrystals that diffract at high resolution (up to 1.78 Å) and are robust in various experimental setups. The presence of four calcium ions and a catalytic zinc ion contributes to its pronounced stability [19–21]. Moreover, it can be used as a surrogate model for other zinc-dependent metalloproteases such as neprilysin (NEP), sharing conserved active-site features and substrate specificity [22].

These features make thermolysin highly suitable as both a routine standard sample and a model for mechanistic and inhibitor studies.

1.3. *Glucose Isomerase (Xylose Isomerase)*

Glucose isomerase, also known as xylose isomerase, from *Streptomyces rubiginosus* (43.3 kDa), is another standard sample in serial crystallography. Large-scale industrial production and commercial availability as a purified enzyme allow for the straightforward preparation of homogeneous microcrystals, which typically diffract to around 2 Å.

The protein has two metal-binding sites for magnesium ions in its active site, which can be replaced with other divalent metals [23]. In a study byKovalesky et al. (2010) [23], Cd^{2+} and Ni^{2+} were used as alternative metals to capture different reaction points, as their binding allows sugar binding but inhibits the catalytic reaction before and after the sugar ring-opening step, respectively. This makes glucose isomerase an interesting standard sample for the establishment of novel time-resolved mixing approaches. It is also notable that the activity of the protein is either inhibited or activated upon binding with different metals, depending on their respective ionic radii [24,25].

The protein has been used to study various viscous media as injection matrices [26–30], to test one fixed-target sample holder [31], and for fixed-target pink-beam serial synchrotron crystallography [32]. Additionally, it has been employed to establish mixing via a liquid application method using a spit robot [33,34] and on-chip crystallization [35].

1.4. Proteinase K

Proteinase K from *Engyodontium album* (formerly known as *Tritirachium album*) is a model protein for subtilisin-like serine proteases, and its function is widely studied. It is a small exo- and endoprotease (29.5 kDa) and is routinely used in molecular biology during nucleic acid isolation.

The protein can be purchased as lyophilized powder and microcrystals that diffract up to ~1.8 Å can be obtained within hours.

In serial crystallography, proteinase K has contributed to testing high-speed data acquisition (e.g., kilohertz frame rates) [12], on-chip crystallization [15,35], pink-beam experiments [36,37], SIRAS (Single Isomorphous Replacement with Anomalous Scattering) phasing [38] and other innovations across both XFEL-based and synchrotron-based platforms [39,40].

It has also been used in the development of a new sample delivery methods, including the implementation of circular motion in microfluidic sample plates [41] and in the advancement of fixed-target systems [42].

1.5. Trypsin

Trypsin is a well-characterized serine protease most commonly sourced from porcine or bovine pancreas and is widely employed as a standard in serial crystallography. Trypsin's structural stability and predictable enzymatic specificity have made it valuable for benchmarking in situ data collection methods [43], high-throughput ligand screening [44], and the evaluation of new droplet microfluidic device for microcrystal production [45]. It was also used for the commissioning of a high-speed piezo-driven goniometer [46].

In addition, trypsin and its variants are used as model systems for testing protease inhibitors and for structural studies relevant to drug discovery. The range of comparative data and reproducible diffraction quality strengthens trypsin's position as a reliable reference sample for the development and standardization of serial crystallography techniques.

1.6. Myoglobin

Among proteins studied by serial crystallography, myoglobin holds a special place as both a scientific milestone and a versatile experimental standard. Famous as the first protein whose structure was solved at atomic resolution, it continues to serve as a bridge between classical crystallography and modern serial methods [47,48].

The heme-containing globular protein is especially valuable for time-resolved studies, as its ligand-binding and photodissociation reactions can be triggered and monitored within microcrystals [48,49]. This property has made myoglobin a key model for investigating ultrafast structural dynamics using pump-probe serial crystallography experiments, as well as for static structure determination at room temperature [50–52].

Microcrystallization protocols for myoglobin are well established, though crystallization behavior varies significantly between species [48,51,53]. Therefore, crystallization conditions often need to be optimized for the specific myoglobin variant used. Despite these differences, batch crystallization of myoglobin can reliably produce large quantities of microcrystals suitable for serial experiments, supporting high-throughput data collection and compatibility with diverse sample delivery methods.

Beyond its technical merits, myoglobin's rich history and photoreactivity have made it a preferred system for demonstrating new experimental approaches and training researchers in serial crystallography. Its continued use reflects both its scientific importance and practical value in advancing structural biology.

1.7. GFP and Its Derivatives

Green fluorescent protein (GFP) from the jellyfish *Aequorea victoria* (~27 kDa) is widely used in molecular biology, particularly in fluorescent imaging and as an intracellular sensor [54]. Numerous engineered variants of GFP have been developed, each with distinct properties that expand its utility across various applications [55–57].

GFP and its derivatives are relatively hydrophobic, exhibit considerable thermostability, and crystallize readily to produce well-diffracting crystals. Among these, the reversibly photoswitchable variant rsEGFP2 has been used to study ultrafast structural changes occurring on timescales from picoseconds to milliseconds [58–60].

Another derivative, eGFP, has also been featured in studies describing the production and handling of microcrystals for serial crystallography [61].

Another engineered GFP variant, iq-mEmerald, was developed as an intracellular metal sensor [62]. This synthetic derivative includes a metal-binding site engineered near the chromophore by substituting three surface residues with histidine (H147, H202, and H204). Transition metals bind to this site, modulating the fluorescence in a concentration- and metal-dependent manner. Specifically, Co^{2+}, Ni^{2+}, and Cu^{2+} ions quench fluorescence, whereas mixing with Zn^{2+} results in increased fluorescence. Notably, copper binding induces up to 80% fluorescence quenching, with an affinity (K_a) of approximately 0.2 μM [62]. These properties make iq-mEmerald particularly useful for visualizing mixing efficiency and studying diffusion processes in time-resolved serial crystallography. For example, it has been employed to characterize a newly designed mix-and-inject high-viscosity extruder, taking advantage of its fluorescence sensitivity to monitor real-time mixing dynamics within the nozzle [63]. In such experiments, mixing efficiency depends on multiple factors, including the viscosity of the carrier matrix, the crystal morphology, and the arrangement of protein molecules within the crystal lattice.

1.8. PYP

Photoactive Yellow Protein (PYP) is a well-established model system in serial crystallography, particularly for time-resolved studies [64–68]. Its suitability comes from a thoroughly characterized photocycle, consistent crystallization behavior, and compatibility with pump-probe experimental setups. PYP is a small (~14 kDa), water-soluble photoreceptor protein that undergoes a trans-to-cis isomerization of its covalently bound *para*-coumaric acid (*p*CA) chromophore upon blue-light excitation, initiating a series of structural intermediates spanning femtoseconds to seconds [66,69–71].

PYP is readily crystallized in batch, producing homogeneous microcrystals suitable for various delivery methods, including liquid jets [66,67] and fixed targets [72]. It has played a key role in advancing time-resolved serial femtosecond crystallography (TR-SFX) at XFELs, where its full photocycle has been resolved at atomic resolution (up to 1.46 Å) under room-temperature conditions [66]. These studies confirm that SFX can reveal detailed structural changes with minimal radiation damage, even at high X-ray doses.

Comparative investigations of PYP dynamics in crystalline and solution environments have provided valuable insights into how the crystal lattice can influence reaction pathways and functional mechanisms [64,66,68,73,74].

Beyond its biological significance, PYP is extensively used to develop and validate sample delivery systems, data processing pipelines, and experimental protocols in serial crystallography. Studies with PYP have substantially advanced our understanding of light-induced signal transduction and protein dynamics.

1.9. Thaumatin

Thaumatin from *Thaumatococcus daniellii* is one of the most widely used standard samples in serial crystallography (SX), including both synchrotrons [39,75,76] and XFELs [30,77–79]. Its popularity is due to its robust and reproducible crystallization, well-characterized structure, and adaptability to a wide range of sample delivery and data collection methods. Thaumatin readily forms high-quality microcrystals in batch, allowing the preparation of dense suspensions ideal for serial experiments.

Various sample delivery approaches have been tested with thaumatin, including fixed-target chips (such as polymer-based membranes and silicon chips) [76,80], liquid jets [77], and HVE [30,39]. Direct on-chip crystallization allows efficient in situ serial data collection with minimal sample handling and low background scattering, and supports ligand soaking and high-throughput screening [15,81].

Because of its reproducibility and the abundance of comparative data available, thaumatin is routinely used for beamline commissioning, validation of sample delivery systems, and testing new data collection strategies. Well-established protocols for batch microcrystallization have been successfully adapted to various delivery systems, further supporting its role as a versatile standard in serial crystallography.

1.10. Granulovirus

Granulovirus occlusion bodies (OBs) are a distinctive standard sample in serial crystallography, particularly valuable for experiments that test the limits of crystal size and radiation dose tolerance. These naturally occurring protein nanocrystals, produced in vivo by insect viruses such as *Cydia pomonella* granulovirus (CpGV), encapsulate the viral protein granulin within a crystalline matrix, forming highly homogeneous particles typically measuring around 100 × 100 × 300 nm [82]. Each OB contains approximately 9000 unit cells, with volumes less than 0.016 μm^3, making them among the smallest biological crystals used in serial crystallography [83–85].

Their uniformity and narrow size distribution make granulovirus OBs ideal standard samples for SFX experiments. SFX data collection at XFELs has enabled the determination of granulin structure from these OBs at atomic resolution at very high X-ray doses. The use of femtosecond pulses minimizes radiation damage, allowing high-quality data collection from these sensitive nanocrystals at room temperature [83]. Due to their reproducibility, challenging size regime, and high hit rates, granulovirus OBs have become a standard sample for evaluating and comparing different serial crystallography platforms, including both XFEL-based SFX and serial electron diffraction (SerialED) [85].

In this review, we focus on four representative standard samples: lysozyme, myoglobin, iq-mEmerald, and PYP. Each was selected for its distinct characteristics and relevance to different aspects of SX experiments. Our selection criteria were based on the diverse physicochemical properties, functional relevance, and established or emerging roles of these proteins in proof-of-concept and method development studies.

Lysozyme serves as the classical standard for protein crystallography, thanks to its small size (14.3 kDa), commercial availability, and remarkable ability to crystallize under a wide range of conditions. The robustness of lysozyme makes it an ideal standard sample for testing sample delivery methods, crystallization replicability, injector compatibility, and data collection protocols [7]. In addition, its extensive literature coverage allows users to calibrate new workflows and compare performance directly across instruments and facilities.

Table 1. Summary of key properties for serial crystallography standard proteins, including UniProt identifiers, molecular weights, primary experimental applications, references, micro-crystallization conditions (only for proteins for which protocols are not reported in this article), and images of tertiary structure. Ligands and chromophores (colored) are being depicted as sticks or spheres (C = grey; Fe = orange; O = red; Zn = dark blue; N = navy blue; Ca = neon green; Mg = lime green) [1]. Due to the breadth of available studies, only selected references are cited for each protein. * Protein buffer condition not reported in references. Tertiary structures display the secondary structure of one monomer, respectively.

Protein (Uniprot ID)	Size [kDa]	Major Use	References [1]	Reported Micro-Crystallization Conditions	Tertiary Structure
PYP (P16133)	13.87	Proof of principle Time-resolved study Sample delivery development	[64,65,67] [65,66,68] [72]	This work	 PDB ID 6P5G
Lysozyme (P00698)	14.31	Instrument commissioning Proof of principle Sample delivery development	[33,86–90] [90,91] [8,10,17,26,33,92–94]	This work	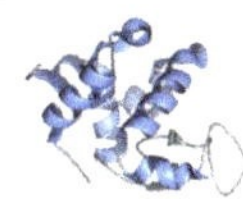 PDB ID 9I6N
Myoglobin (P68082, P02185)	16.95	Instrument commissioning Time-resolved study Proof of principle Sample delivery development	[52,95,96] [48] [50,52,97] [50,51]	This work	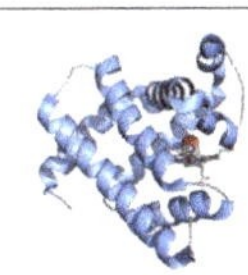 PDB ID 8BKH
Thaumatin (P02883)	22.21	Instrument commissioning Proof of principle Sample delivery development	[39] [75,77–79] [15,30,76,80,81]	100 mg/mL protein in ddH$_2$O + 1.6 M sodium potassium tartrate [79]	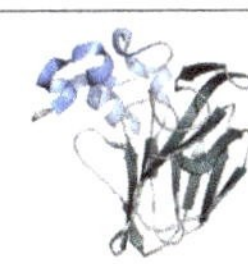 PDB ID 9FTS
Trypsin (P00760)	23.56	Sample delivery development	[43–46]	30 mg/mL protein in 25 mM HEPES (pH 7.0), 5 mM CaCl$_2$ + 100 mM Tris (pH 8.5), 30%(*w/v*) PEG 3350, 200 mM Li$_2$SO$_4$ [43] (30 mg/mL protein + 10 mg/mL benzamidine in 20 mM HEPES (pH 7.0), 10 mM CaCl$_2$) + (20% PEG 8000, 200 mM (NH$_4$)$_2$SO$_4$, 100 mM Bis-Tris) [44] * (65 mg/mL protein + benzamidine in 3 mM CaCl$_2$) + (11–14%(*w/v*) PEG 4000, 15% ethylene glycol, 200 mM SiSO$_4$, 100 mM MES (pH 6.5)) [45] (15 mg/mL protein + 5 mg/mL benzamidine in 10 mM CaCl$_2$, 20 mM HEPES (pH 7.0), 3.75% PEG 3350, 5% glycerol) + (15% PEG3350, 20% glycerol) (hanging drop) [46]	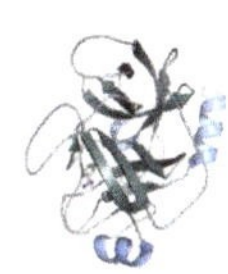 PDB ID 7WA0
iq-mEmerald	27.1	Sample delivery development	[63]	This work	 PDB ID 4KW4

Table 1. *Cont.*

Protein (Uniprot ID)	Size [kDa]	Major Use	References [1]	Reported Micro-Crystallization Conditions	Tertiary Structure
rsEGFP2	26.9	Instrument commissioning Time-resolved study	[98] [59,60,99]	20 mg/mL protein in 2M $(NH_4)_2SO_4$, 20 mM NaCl, 120 mM HEPEs (pH 8.0) (seeding) [59] 20–24 mg/mL protein in 75 mM HEPES (pH 8.0), 20 mM NaCl, 1.1–1.3 M $(NH_4)_2SO_4$ (seeding) [99]	 PDB ID 5O89
Proteinase K (P06873)	29.1	Instrument commissioning Proof of principle Sample delivery development	[36,37,39,40] [38] [15,35,41,42]	40 mg/mL protein in 20 mM MES (pH 6.5) + 100 mM MES (pH 6.5), 500 mM $NaNO_3$, 100 mM $CaCl_2$ [39]	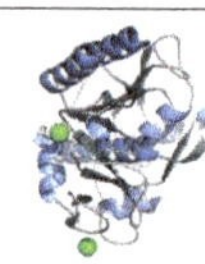 PDB ID 9FTX
Thermolysin (P00800)	34.86	Proof of principle Sample delivery development	[100,101] [15–18,44,46]	22.5 mg/mL protein in 100 mM MES (pH 6.5) + 10 mM $CaCl_2$, 5% PEG 2000 [15] * 30 mg/mL protein in 50 mM NaOH + 15% (*w/v*) ammonium sulfate [16]. * 42.5 mg/mL protein + 40% PEG 2000 MME, 0.1 M MES (pH 6.5), 5 mM $CaCl_2$ [18] 330 mg/mL protein + 45% DMSO in 50 mM Tris (7.5) + 1.45 M $CaCl_2$ [44]	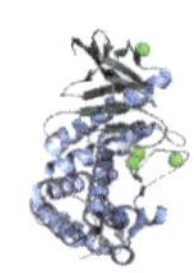 PDB ID 5WR4
Glucose isomerase (P24300)	43.33 (monomer) 173.32 (homo tetramer)	Instrument commissioning Sample delivery development	[32] [26–31,33–35]	33 mg/mL protein in 6 mM Tris (pH 7.0), 0.91 M $(NH_4)_2SO_4$, 1 mM $MgSO_4$ [31] * 80 mg/mL protein + 35%(*w/v*) PEG3350, 0.2 M LiSO4, 10 mM HEPES (pH 7.5) [35]	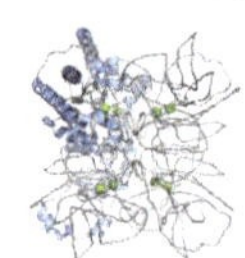 PDB ID 6KD2
Granulovirus (Granulin) (P87577)	29.38 (monomer) 352.56 (homo 12-mer)	Proof of principle Sample delivery development	[83] [84]	Not applicable	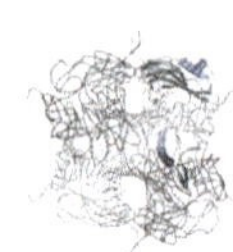 PDB ID 5G0Z

Myoglobin was chosen as a standard sample because it combines well-established crystallization protocols with significant experimental flexibility. Its well-characterized heme cofactor and robust crystal formation allow it to serve as a reliable model for both static and time-resolved serial crystallography experiments. Importantly, myoglobin supports advanced studies of ultrafast structural changes, such as ligand photodissociation, making it highly relevant for pump-probe SFX applications [48]. Additionally, its widespread use across XFEL and synchrotron facilities provides a valuable standard for method development and cross-facility comparisons.

Iq-mEmerald is a variant of the green fluorescent protein designed for enhanced brightness and stability. As a fluorescent marker, it enables direct visualization of microcrystal suspension quality, facilitates monitoring of crystal flow during injection, and supports assessment of mixing with ligands. The use of iq-mEmerald provides a unique supporting tool for optimizing and troubleshooting sample delivery, especially in mixing injections [63].

PYP was selected to represent the class of light-activating proteins, which are highly relevant for time-resolved studies probing ultrafast structural changes. Its well-characterized photocycle and robust crystallization make it an ideal candidate to develop and validate experimental approaches for pump-probe SFX [66,69]. Its inclusion highlights specific challenges and solutions in preparing photosensitive samples and supports the evaluation of light-triggered reaction methodologies.

We present detailed protocols for the preparation of these standard samples specifically tailored for SFX experiments. Compared to SSX (serial synchrotron crystallography), SFX requires large quantities of highly uniform microcrystals, typically in the low micrometer range or smaller, to ensure efficient and reliable sample injection into the XFEL beam. Furthermore, sample homogeneity and filtering are critical to prevent clogging of injectors and to maximize data quality. The crystals must also be stable and well-suspended in compatible carrier media adapted to high-speed injection. Covering all steps from protein expression and purification (if applicable) to crystallization, these protocols provide comprehensive and reproducible preparation techniques and support the broader SFX community in ensuring high-quality, reliable data for both instrument commissioning and experimental validation.

2. Materials and Methods

2.1. *Lysozyme*

2.1.1. Materials

- Lysozyme: Carl Roth GmbH + Co. KG (Karlsruhe, Germany) Art.no. 8259;
- Sodium Acetate (NaOAc): Merck KGaA (Darmstadt, Germany) Art.no. 71183;
- Sodium chloride (NaCl): Carl Roth GmbH + Co. KG Art.no. P029;
- PEG 6000: Merck KGaA Art.no. 81260;
- Monoolein: Nu-Chek Prep (Elysian, MN, USA) Art.no. M-239;
- Gravity filters: CellTrics^{TM}, Sysmex Deutschland GmbH (Norderstedt, Germany) (10 μm filter, Art. No. 04-0042-2314 and 20 μm filter, Art. No. 04-0042-2315);
- Gas-tight glass syringe: Hamilton Bonaduz AG (Bonaduz, Switzerland), Art.no. 202668;
- Syringe coupler:Rigaku Holdings Corporation (Tokyo, Japan), Art.no. EB-LCP-SUNION.

2.1.2. Prepared Solutions

- 0.5 M NaOAc, pH 3.5;
- 5 M NaCl;
- 50% (*w/v*) PEG 6000;
- Crystallization solution: 0.1 M NaOAc, pH 3.5, 5% PEG 6000 (*w/v*), 3.2 M NaCl, 0.2 μm filtered;
- Lysozyme solution: 100 mg/mL in 50 mM NaOAc, pH 3.5, 0.2 μm filtered;
- Storage buffer: 50 mM NaOAc, pH 3.5, 1.7 M NaCl, 0.2 μm filtered.

2.1.3. Crystallization

Lysozyme crystals were generated by vortexing a 1:1 ratio of the prepared lysozyme solution and crystallization solution under controlled temperature conditions. Prior to mixing, both the protein solution and the crystallization solution were equilibrated in a thermoblock. Variations in the mixing protocol, either initiating vortexing before adding the crystallization solution or adding the crystallization solution followed by immediate vortexing, led to differences in crystal size. Therefore, a standardized mixing procedure should be employed to guarantee reproducibility.

2.1.4. Storage Buffer Exchange

Following crystallization, the liquid was exchanged with storage buffer. This was achieved through three rounds of centrifugation. Due to the higher viscosity of the crystallization solution, the first centrifugation was performed at 200× *g* for one minute, while the subsequent two rounds were conducted at 100× *g* for one minute each. After each centrifugation step, the supernatant was removed, and the crystal pellet was resuspended with an equal volume of fresh storage buffer.

2.1.5. Crystal Filtration and Density Adjustment

To remove remaining large particles, the stored crystals were filtered using Nylon mesh gravity filters. A 10 µm filter was used for crystals smaller than 5 µm, while a 20 µm filter was employed for crystals ranging from 5 to 10 µm. The filtered crystals were allowed to settle overnight, and the sedimented volume was measured the following day. Depending on the purpose of this sample, the density (vol % of sedimented crystal) was fixed by removing or adding the storage buffer.

2.1.6. Embedding in LCP

Lysozyme crystals with sizes of 5–7 µm were prepared with aqueous solutions as described above. Prior to embedding the lysozyme crystals in the lipidic cubic phase (LCP), the LCP containing the lysozyme crystal storage buffer was prepared as follows: First, the lysozyme crystal storage solution was diluted with water to 40% of its original concentration, as the undiluted solution disrupted the LCP structure and produced an opaque material. A 40% dilution was found to be the upper limit for forming a transparent LCP using this buffer composition. The diluted solution was then transferred into a gas-tight glass syringe. A second syringe was filled with melted monoolein. The volume ratio of the two solutions was fixed at 2:5 for the diluted lysozyme crystal storage solution and monoolein, respectively. The two syringes were connected using a syringe coupler, and the solutions were mixed by moving the plungers back and forth. Once the mixture became transparent, the entire sample was transferred into one syringe, and the other syringe was removed.

The filled syringe was then connected to a third syringe containing a lysozyme crystal pellet of 10% (*v/v*) relative to the prepared LCP in the other syringe. The crystals were mixed into the prepared LCP by moving the plunger until the pellet was evenly distributed throughout the syringe.

2.2. Myoglobin

2.2.1. Materials

- Myoglobin (equine skeletal muscle): Merck KGaA Art.no. M0630;
- Ammonium sulfate ($(NH_4)_2SO_4$): Carl Roth GmbH Art.no. 9212.1;
- Tris (Tris-(hydroxymethyl)-amino methane): Carl Roth GmbH Art.no. 5429.3;
- Sodium dithionite: Merck KGaA Art.no. 71699;
- 40 µm frit filter: JR-1100-40P, Valco Instrument Co. Inc. (Houston, TX, USA);
- PreColumn: A-355, IDEX Health & Science LLC (Rohnert Park, CA, USA);
- Luer adapter: P-642, IDEX Health & Science LLC.

2.2.2. Prepared Solutions

- 4 M $(NH_4)_2SO_4$, 0.2 µm filtered;
- 50 mM Tris buffer, pH 7.5, 0.2 µm filtered;
- 0.5 M sodium dithionite in degassed 3.3 M $(NH_4)_2SO_4$, prepared in a glove box (GS MEGA 4).

2.2.3. Crystallization

Myoglobin powder was transferred to a 5 mL centrifuge tube, and its weight was measured. Due to the volume limit of the tube, 20–30 mg of myoglobin is the optimal amount. The powder was then dissolved in 50 mM Tris buffer (pH 7.5) to achieve a final concentration of 23.6% (*w/v*). To ensure complete dissolution, the tube was vortexed thoroughly, and a brief centrifugation step was performed to collect all liquid at the bottom.

While the myoglobin solution was being vortexed, the 4 M $(NH_4)_2SO_4$ solution was added dropwise. The total volume of $(NH_4)_2SO_4$ solution added was 4.55 times the volume of the Tris buffer used for myoglobin dissolution. During this process, the initially transparent brown solution became cloudy, indicating the onset of nucleation. Vortexing continued for an additional 30 s, and the mixture was left undisturbed at room temperature overnight.

2.2.4. Crystal Filtration

On the following day, crystal formation was inspected, and the resulting crystal slurry was filtered through a 40 μm frit filter and placed in a pre-column, which was connected to a Luer adapter and a syringe, in order to remove undesired aggregates and to disassemble clustered crystals. For liquid jet sample delivery, the crystal suspension was diluted 2-fold relative to the original crystallization volume using 3.3 M $(NH_4)_2SO_4$.

2.2.5. Deoxygenation

Prior to placing the samples into the glove box, a 3.3 M $(NH_4)_2SO_4$ solution was degassed using a vacuum pump. The tubes containing crystals and the degassed 3.3 M $(NH_4)_2SO_4$ were transferred into the glove box with an oxygen concentration of less than 0.5 ppm. A 0.5 M sodium dithionite solution was added to the myoglobin crystals to reach a final concentration of 3 mM, and the reduction reaction was allowed to proceed for the next 3 min. The reaction was monitored through a color change from dark brown to light red. Once the reaction was complete, sodium dithionite was removed by buffer exchange using centrifugation. The solution was centrifuged at $500\times$ *g* for 10 min, and the supernatant was removed. The crystal pellets were then resuspended in an equal volume of degassed 3.3 M $(NH_4)_2SO_4$ solution. This process was repeated five times to remove sodium dithionite completely.

2.3. Iq-mEmerald

2.3.1. Materials

- Plasmid: iq-mEmerald expression vector pET17b-iq-mEmerald without using the fusion tag. The vector was purchased from Biocat GmbH (Heidelberg, Germany) using the amino acid sequence taken from the Fluorescent Protein Data Base (https://www.fpbase.org/protein/7G47U/ (accessed on 10 September 2025)) and pET17b as vector backbone.
- Recipient cell line: *Escherichia coli* BL21(DE3) (Thermo Fisher Scientific Inc. (Waltham, MA, USA), Art.no. EC0114).
- Glassware: 5 L flasks.
- Seed Beads: Jena Bioscience GmbH (Jena, Germany), Art.no. CO-501.
- Gravity filter: CellTrics™ 30 μm, Sysmex Deutschland GmbH, Art. Nr. 04-0042-2316.
- LB-medium: Carl Roth GmbH Art.no. 6673.4.
- Ampicillin: Carl Roth GmbH Art.no. K029.2.
- IPTG (isopropyl β-D-thiogalactopyranoside): Carl Roth GmbH Art.no. 2316.5.
- Tris (Tris-(hydroxymethyl)-amino methane): Carl Roth GmbH Art.no. 5429.3.
- Sodium chloride (NaCl): Carl Roth GmbH Art.no. P029.
- Ammonium sulfate ($(NH_4)_2SO_4$): Carl Roth GmbH Art.no. 9212.1.

- Ethanol (99.9%): Merck KGaA Art.no. 1.00983.
- Hydrophobic interaction column (HIC), e.g., HiPrep Phenyl FF (High Sub) 16/10 Cytiva, Marlborough, MA, USA Art. Nr 28936545).
- 10 kDa cut-off concentrator: Merck KGaA Art.no. UFC9010.

2.3.2. Equipment

- Incubation shaker: Eppendorf SE (Hamburg, Germany) New Brunswick™ Innova®44/44R Shaker;
- FPLC: Cytiva ÄKTA pure™ chromatography system.

2.3.3. Prepared Solutions

- Antibiotic: 100 mg/mL Ampicillin stock solution in ethanol;
- 1 M IPTG (isopropyl β-D-thiogalactopyranoside); 0.2 μm filtered;
- Buffer A (Lysis Buffer): 20 mM Tris, pH 7.8, 150 mM NaCl; 0.2 μm filtered;
- Buffer B (HIC start buffer): 20 mM Tris, pH 7.8, 20% $(NH_4)_2SO_4$ saturation, 0.2 μm filtered;
- Buffer C (HIC elution buffer): 20 mM Tris, pH 7.8, 0.2 μm filtered;
- 5 M NaCl, 0.2 μm filtered;
- 70% $(NH_4)_2SO_4$ saturated solution;
- Crystallization Buffers: 50 mM Tris, pH 8.0, 1.5–3 M $(NH_4)_2SO_4$ concentrations in 0.1 M increments, 0.2 μm filtered.

2.3.4. Expression

One colony of *Escherichia coli* BL21 (DE3) pET17b-iq-mEmerald was inoculated in 50 mL of LB medium supplemented with 100 μg/mL Ampicillin and incubated at 37 °C at 180 rpm for 16 h. The following day, the culture was diluted to an OD_{600} of 0.1 in 1 L fresh LB medium containing 100 μg/mL Ampicillin and grown at 37 °C, 180 rpm, until the OD_{600} reached 0.6. Protein expression was induced by the addition of 0.5 mM IPTG, and the culture was incubated at 18 °C for 16 h with shaking at 180 rpm. Cells were harvested by centrifugation at 8000× *g* for 15 min to 1 h at 4 °C, and pellets were stored at −20 °C until further use.

2.3.5. Purification

The purification strategy was adapted from Samarkina et al. (2009) [102].

The cell pellet was resuspended in Buffer A (10 mL Buffer A per 1 g of cell pellet). The cells were lysed by sonication with 50% amplitude and 10 s on 10 s off cycles until 400 J were reached. After centrifugation of the cell lysate at 7197× *g* for 15 min at RT, the supernatant was transferred to a new tube and incubated at 65 °C for 15 min in a thermocycler or water bath. The supernatant turned cloudy and spun down for 15 min at RT and 7197× *g*. Ten milliliters of the supernatant were mixed rapidly by brief vortexing with 3 mL of 5 M NaCl and 23.3 mL of saturated $(NH_4)_2SO_4$ (pH 7.8), respectively. Twelve milliliters of 99.9% ethanol were added instantly, and tubes were vortexed for 30 s. After that, samples were centrifuged for 7 min at RT and 3000× *g*. Iq-mEmerald is present in the organic phase, which was carefully removed and transferred to a new tube. The protein-containing fraction was diluted to 20% $(NH_4)_2SO_4$ saturated solution in 20 mM Tris, pH 7.8. Before subjecting the sample to a HIC equilibrated with Buffer B, it was filtered through a 0.2 μm syringe filter. Iq-mEmerald was eluted over 20 column volumes of Buffer C. The fractions containing iq-mEmerald were concentrated with a 10 kDa cut-off concentrator to 50 mg/mL, flash frozen in liquid nitrogen and stored at −80 °C.

2.3.6. Crystallization

After thawing, iq-mEmerald samples were filtered with a 0.2 µm filter or centrifuged at RT for 10 min at 20,817× *g*.

Seedstock

To prepare a large, high-concentration seedstock, 500 µL iq-mEmerald was mixed with 500 µL of 3 M $(NH_4)_2SO_4$ and vortexed immediately for 30 s. The next day, inhomogeneous iq-mEmerald crystals appeared. Approximately 20 to 30 small glass beads were added to a 1.5 mL reaction tube. The tube was vortexed vigorously for 5 min. The slurry was cooled down to room temperature, and vortexing was repeated until no crystals were visible under a stereo microscope (total 20 min process time).

Needle Crystals

To grow needle crystals with a size of 1 × 1 × 12 µm, 500 µL protein solution (50 mg/mL) was mixed with 1000 µL of 2 M $(NH_4)_2SO_4$, 50 mM Tris (pH 8.0), as well as 5 µL seedstock. The sample was vortexed for 30 s and then incubated at RT. Crystals grew overnight to their final size.

Cubic Crystals

To grow cubic crystals with a size of 5 × 5 × 5 µm, 500 µL protein solution (50 mg/mL) was mixed with 1000 µL of 3 M $(NH_4)_2SO_4$, 50 mM Tris (pH 8.0), as well as 5 µL seedstock and vortexed for 30 sec. After incubation for 1 min, 500 µL of 2.5 M $(NH_4)_2SO_4$ was added. Crystals grew overnight to their final size.

All crystal slurries were filtered with a 20 µm gravity filter and stored at RT.

2.3.7. Embedding in LCP

Iq-mEmerald crystals of various sizes and shapes, as prepared above, were suitable for embedding in LCP. For this, a transparent LCP matrix was first prepared by mixing monoolein and the iq-mEmerald crystal storage buffer, 50 mM Tris (pH 8.0) and 1.5–3 M $(NH_4)_2SO_4$, in a 7:3 volume ratio using two gas-tight glass syringes connected by a coupler. The crystal storage buffer was transferred into one gas-tight glass syringe, and a second syringe was loaded with melted monoolein. The two syringes were connected and mixed thoroughly by repeated plunger exchange until a clear, homogeneous LCP formed. The mixture was combined into one syringe.

To incorporate the crystals, a third syringe containing an iq-mEmerald crystal pellet (10% *v/v* relative to the prepared LCP) was connected to the LCP-containing syringe. The crystals were embedded by mixing with the LCP until they were evenly distributed throughout the matrix, producing a final preparation suitable for high-viscosity injection.

2.4. PYP (Photoactive Yellow Protein)

2.4.1. Materials

- Plasmid: pET-M11 [103]. The expression vector containing the codon-optimized gene sequence for PYP was purchased from Biocat GmbH (Heidelberg, Germany) using the full PYP sequence from UniprotKB: P16113.
- Recipient cell line: *Escherichia coli* Rosetta(DE3), Merck KGaA, Art.no. 70954-3.
- Glassware: 5 L flasks.
- Seed Beads: Jena Bioscience GmbH, Art.no. CO-501.
- PD-10 Buffer exchange columns: Cytiva, Marlborough, MA, USA, Art. No. 17085101.
- Gravity filter: CellTrics™ 30 µm, Sysmex Deutschland GmbH, Art. Nr. 04-0042-2316.
- NiNTA: Thermo Fisher Scientific Art.no. A50586.
- Gravity column: Carl Roth GmbH Art. No. 1518.1.

- LB-medium: Carl Roth GmbH Art.no. 6673.4.
- Kanamycin: Carl Roth GmbH Art.no. T832.4.
- Chloramphenicol: Thermo Scientific Chemicals Art.no. B20841.22.
- HEPES (*N*-2-Hydroxyethylpiperazine-*N′*-2-ethane sulphonic acid): Carl Roth GmbH Art.no. 9105.3.
- Sodium chloride (NaCl): Carl Roth GmbH Art.no. P029.
- Imidazole: Carl Roth GmbH Art.no. 3899.3.
- Tris (Tris-(hydroxymethyl)-amino methane): Carl Roth GmbH Art.no. 5429.3.
- Tri-sodium citrate dihydrate: Carl Roth GmbH Art.no. 3580.1.
- Citric acid: Carl Roth GmbH Art.no. 7624.1.
- *p*-Coumaric acid: Merck KGaA, Art.no. C9008.
- *N,N′*-Dicyclohexylcarbodiimide (DCC): Merck KGaA, Art.no. D80002.
- *N,N*-Dimethylformamide (DMF): Merck KGaA, Art.no. 227056.
- Sodium malonate (Na-malonate): Merck KGaA, Art.no. M4795.
- Beta-mercaptoethanol: Carl Roth GmbH Art. No. 4227.3.
- Glycerol: Carl Roth GmbH Art. No. 3783.2.
- 3 kDa cut-off concentrator: Merck KGaA Art.no. UFC9003.

2.4.2. Equipment

- Incubation shaker: New Brunswick™ Innova®44/44R Shaker;
- Anion exchange column: HiPrep Q HP 16/10 Cytiva Art.no. 29018182;
- FPLC: Cytiva ÄKTA pure™ chromatography system.

2.4.3. Prepared Solutions

- Kanamycin stock solution (100 mg/mL in ddH_2O);
- Chloramphenicol stock solution (34 mg/mL in ethanol);
- Buffer A (Lysis Buffer): 20 mM HEPES, pH 7.4, 200 mM NaCl, 5 mM Imidazole, 0.2 μm filtered;
- Buffer B (Wash buffer): 20 mM HEPES, pH 7.4, 200 mM NaCl, 10 mM Imidazole, 0.2 μm filtered;
- Buffer C (Elution buffer): 20 mM HEPES, pH 7.4, 200 mM NaCl, 300 mM Imidazole, 0.2 μm filtered;
- Buffer D (TEV protease reaction buffer): 20 mM HEPES, pH 7.4, 200 mM NaCl, 0.2 μm filtered;
- Buffer E (Anion exchange start buffer): 25 mM Tris, 0.2 μm filtered;
- Buffer F (Anion exchange elution buffer): 25 mM Tris, 1 M NaCl, 0.2 μm filtered;
- Buffer G (Storage buffer): 50 mM Citrate Buffer, pH 6.0, 0.2 μm filtered;
- Buffer H (Crystallization buffer): 3.7 M sodium malonate, pH 7, 0.2 μm filtered;
- Tobacco Etch Virus (TEV) protease: prepared following the protocol from Berg et al. (2006) [104], in 50 mM Tris (pH 8.0), 200 mM NaCl, 5 mM beta-mercaptoethanol, and 10% (*v/v*) glycerol.

2.4.4. Expression

The purification procedure was adapted from Schmidt et al. (2019) [105], and chromophore production was derived from Kim et al. (2013) [106].

One colony of *Escherichia coli* Rosetta(DE3) pET-M11 (Rosetta(DE3)-PYP) was inoculated in 200 mL LB (containing 50 μg/mL Kanamycin and 34 μg/mL Chloramphenicol) in a 1000 mL flask and incubated overnight at 37 °C and 160 rpm. Twelve 5 L flasks were each filled with 1 L of LB medium containing Kanamycin (50 μg/mL). Fifteen milliliters of the preculture were added to each flask and incubated for approximately 1 h 40 min at 37 °C and 160 rpm. After the OD_{600} reached 0.5, 1 mM IPTG was added, and the cells were incubated for 16 h at 18 °C and 160 rpm. The cells were harvested by centrifugation at 8000× *g* for at least 15 min at 4 °C. The cell pellet from a 1 L culture was transferred to a 50 mL Falcon tube and stored at −20 °C.

2.4.5. *p*-Coumaric Anhydride (*p*CA) Synthesis

The chemicals used are hazardous or toxic, so all tasks were performed under a fume hood and appropriate PPE was worn. DCC was dissolved in 50 mL DMF to a final concentration of 1 mM (2.7 g). Separately, *p*-Coumaric acid was dissolved in 50 mL DMF to a final concentration of 0.77 mM (2.55 g). Both solutions were stirred separately at 4 °C until fully dissolved. Once dissolution was complete, the two solutions were combined and stirred overnight at 4 °C. The synthesis can be upscaled as long as the molar ratio of 1:3:1 is kept. The next day, a light-yellow solution with visible white precipitate was obtained. The solution was centrifuged at 8000× *g* for 1 h at 4 °C. The supernatant was then aliquoted to 20 mL and stored at −80 °C.

2.4.6. Purification

Four cell pellets (4 L cell culture) were used per purification. The cells were resuspended in 40 mL per liter of cell culture of Buffer A. The suspension was quite viscous and slimy. The cells were lysed by sonication with 25% amplitude for 3 min (30 s on, 1 min rest), keeping the suspension on ice. The lysate was then centrifuged at 4 °C and 8000× *g* for at least 30 min. The supernatant was transferred into a beaker and stirred slowly. While the cells were being lysed, an aliquot of *p*CA (20 mL) was taken out from the freezer and centrifuged at 7000× *g* for 30 min at RT, to get rid of any remaining precipitate. Twenty milliliters of *p*CA (5 mL per 1 L of cell culture) were added *dropwise* to the cell lysate. First, some precipitate appeared, then the solution turned bright yellow. The solution was stirred gently at 50 rpm for 2 h at RT. To remove the precipitate, the solution was centrifuged at 8000× *g* for 1 h at 4 °C.

The supernatant was loaded onto a 10 mL pre-charged NiNTA resin column (equilibrated with buffer A). The column was then washed with 5 column volumes (CV) buffer A, followed by 5 CV buffer B. The bound protein was eluted with buffer C until the eluate no longer appeared yellow. To remove the imidazole, the eluate was buffer exchanged to buffer D with a PD-10 column. The protein concentration was determined by measuring the absorption at 446 nm, using the equation: conc. (mg/mL) = $OD_{446}/45{,}000 \times 14{,}700 \times$ dilution factor.

TEV protease was added to the purified protein solution with a 1:20 molar ratio (protease/protein), and the solution was gently shaken overnight at 4 °C.

After digestion, 1–2 mL of NiNTA resin in a gravity column was equilibrated with buffer D, and the TEV protease reaction solution was subjected to the column. The flow through, containing the cleaved protein, was collected. The column was washed with buffer D until no yellow fractions were eluted. The uncut protein was eluted using buffer C. The cleaved PYP was concentrated and buffer-exchanged to buffer E using a PD-10 column.

The protein was then further purified using an anion exchange column and an FPLC system. The column was pre-equilibrated with buffer E before the sample was applied. The protein was eluted at 10% buffer F. The absorption was monitored at 280 nm and 446 nm and the fractions containing protein, where the A_{446}/A_{280} ratio was higher than two, were collected.

The protein-containing fractions were combined, and the buffer was exchanged to buffer G using a PD-10 column. The protein was further concentrated to 100 mg/mL, sterile filtered, and flash frozen in liquid nitrogen for storage at −80 °C.

2.4.7. Crystallization

To produce seedstock, 12 μL PYP (100 mg/mL) was mixed with 127 μL buffer H. After combining both solutions, the mixture was stirred slowly overnight. The resulting crystallization slurry was inhomogeneous, containing big and small crystals (Figure 6). The crystals were crushed with seed beads and vigorous vortexing. The seedstock was then diluted 1:3 with buffer H.

The crystal size can be modulated by varying the volume of added seedstock. For the preparation of 2–3 μm sized crystals, 20 μL protein solution (100 mg/mL) was mixed with 86 μL buffer H (final concentration: 3 M Na-Malonate) and 0.5 μL diluted seedstock in a centrifuge tube. The mixture was instantly vortexed and incubated at RT. Crystals appeared quickly and were fully matured within 1 h. Prior to use, the crystals were filtered with a 30 μm gravity filter.

3. Results

3.1. *Lysozyme*

Lysozyme is a widely used standard sample in protein crystallography due to its availability, reproducibility, and excellent diffraction quality [7–10]. Lysozyme crystals serve as a benchmark for optimizing data collection strategies and refining crystallographic methodologies. This report presents an optimized protocol for lysozyme crystallization, final preparation for beamtime, and embedding in lipid cubic phase (LCP) to generate high-quality samples suitable for various applications. Additionally, the protocol enables controlled modification of the lysozyme crystals.

One important consideration is that the protocol for preparing lysozyme crystals of the desired size should be optimized whenever new solutions (lysozyme and/or crystallization solution) are prepared, as it is highly sensitive to any changes. Initial crystallization can be conducted at room temperature, with subsequent adjustments made based on the observed crystal size. As shown in Figure 1, higher temperatures (A, at 23 °C) produced larger crystals (~14 μm), while lower crystallization temperatures (B, at 17 °C) resulted in smaller crystals (~5–7 μm). The smaller crystals formed almost instantly, whereas the larger ones took up to 10 s to form. The initial optimization volume was 500 μL + 500 μL, which was later scaled up to 1.5 mL + 1.5 mL in a 5 mL centrifuge tube.

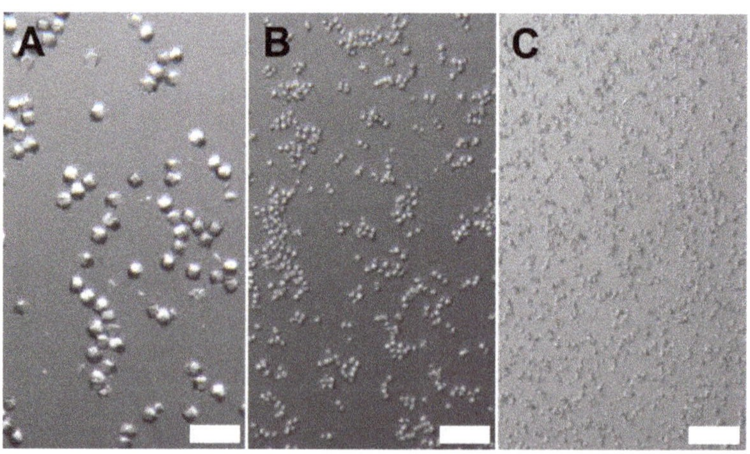

Figure 1. Microscope images of lysozyme crystals. (**A**) Produced at 23 °C (~14 μm); (**B**) produced at 17 °C (~5–7 μm) with vortexing first, then addition of crystallization solution; and (**C**) produced at 17 °C (~2–4 μm) with addition of crystallization solution first, then vortexing. Scale bars: 50 μm.

In addition, as described in the Methods section, consistency in the mixing process is critical for controlling crystal size. Figure 1B shows crystals formed by first vortexing the protein solution, followed by the addition of the crystallization solution, whereas Figure 1C shows crystals obtained when the crystallization solution was added first, followed by vortexing. As shown, the crystals in Figure 1C (~2–4 μm) are smaller than the ones in Figure 1B. This is likely due to more immediate nucleation caused by adding the crystallization solution to the protein solution while it remains still.

The final step of sample preparation for beamtime depends on the sample delivery method. One of the most common sample delivery methods at the European XFEL is the liquid jet, using GDVNs (gas dynamic virtual nozzles) or DFFNs (double-flow focus nozzles) [107,108]. In this application, a density of 15–18% (*v/v*) with 2–5 μm sized crystals provided a stable jet (Figure S1A) with a reasonable hit rate [107]. For sample delivery using HVE, the crystals were embedded in LCP, as described in the Methods section (Figure S1B, Round A. et al., in submission). This highlights the versatile use of lysozyme crystals as a standard sample for serial crystallography.

3.2. *Myoglobin*

Myoglobin crystals serve as a critical model system in serial crystallography due to their well-characterized structural properties and their ability to undergo redox reactions [48,109–111]. These crystals provide valuable insights into protein dynamics and ligand interactions, making them a standard sample for evaluating new methodologies in time-resolved crystallographic studies. In the Methods section, a reliable protocol is described for myoglobin crystallization, enabling the production of high-quality samples for standard applications, as well as the deoxygenation process of the crystals for time-resolved studies.

The crystallization process outlined in the Methods section employs an unconventional ratio between the protein and precipitant (ammonium sulfate), which was determined through experimental optimization. Initially, obtaining consistent crystal growth was challenging; however, this optimized ratio reliably produced crystals. Microscopic inspection of the samples after overnight incubation revealed that some crystals formed in clusters, requiring intense filtration prior to use (Figure 2). Filtration of the crystal slurry using a 40 μm frit filter effectively removed large aggregates. Due to the lower density of myoglobin crystals compared to the crystallization solution, the crystals floated in the solution, making the estimation of crystal density difficult. Therefore, we aimed to maintain the final sample volume for the liquid jet at 2 times the original total volume used in crystallization, which resulted in a stable and reliable liquid jet (Figure S1C).

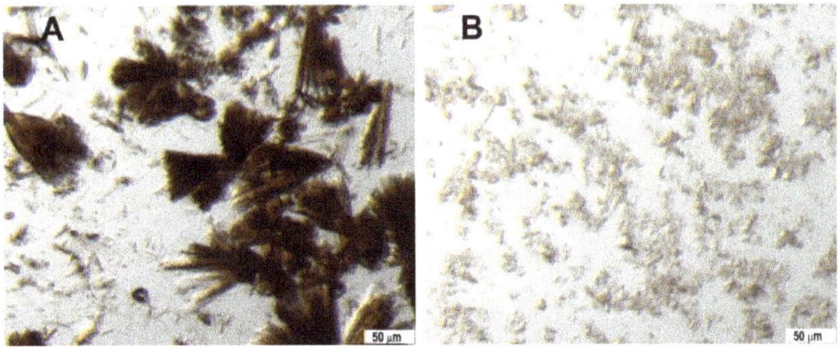

Figure 2. Stereomicroscope images of myoglobin crystals. (**A**) Inhomogeneous large crystals were obtained initially. (**B**) Crystals in (**A**) filtered with 40 μm filter and used for serial crystallography experiment by liquid jet. Scale bars: 50 μm.

Deoxygenation of myoglobin crystals was successfully achieved using the method described here, yielding the deoxy-myoglobin state suitable for time-resolved studies of oxygen binding dynamics. In control experiments, structural analysis of deoxygenated crystals confirmed the absence of oxygen electron density at the heme site. Upon exposure to oxygen during time-resolved SFX experiments, clear electron density corresponding to bound oxygen was observed, confirming the suitability of the prepared crystals for investigating oxygen uptake mechanisms (manuscript in preparation).

The oxidized myoglobin crystals remained stable over time under ambient conditions. The optimized protocol, including crystallization, filtration, and deoxygenation steps, produced crystals suitable for serial crystallography experiments. The reproducibility of the method was confirmed through multiple independent preparations, emphasizing the strength of the protocol.

3.3. *Iq-mEmerald*

There is an ongoing need for improving sample delivery methods for serial crystallography and time-resolved experiments. As we needed a standard sample where mixing could be observed via fluorescence quenching, iq-mEmerald, a GFP derivative, was chosen [62].

The protein can be easily and quickly expressed and purified, leading to >100 mg of pure protein per liter of cell culture. Crystals of different sizes and morphologies can be produced in a reproducible manner by adjusting the concentration of the precipitant and seedstock.

One critical factor for successful mixing and pump-probe experiments is the homogeneity of the microcrystal sample. Therefore, using seedstock is most often the preferred method for producing reproducible, large batches of microcrystals.

A large seedstock was produced from an initially inhomogeneous batch crystallization setup. Several rounds of vortexing were needed to obtain a suitable seedstock (Figure 3). Only after vigorous vortexing for 20 min, a seedstock, not containing larger crystal fragments, was produced. Seedstocks should be prepared in sufficient volumes so that a single batch can be used for an entire experiment. The seedstocks were stored at room temperature for several months without noticeable degradation.

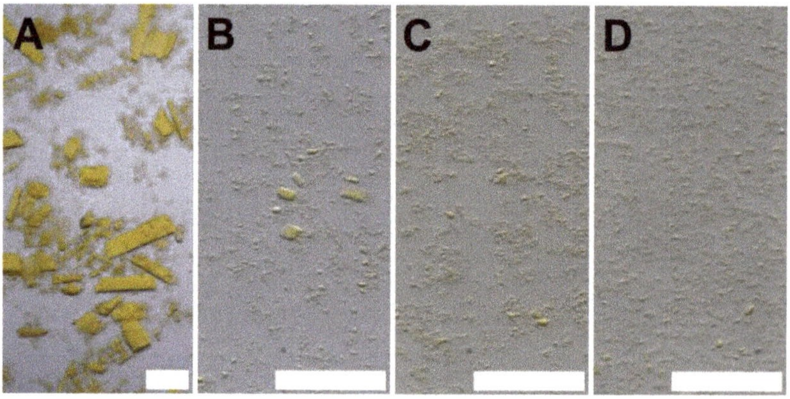

Figure 3. Iq-mEmerald seedstock preparation. A large-scale, homogeneous seedstock was obtained after vortexing a mixture of different-sized crystals with glass beads. (**A**) Inhomogeneous batch crystal suspension; (**B**) after 5 min of vortexing; (**C**) after 10 min of vortexing; (**D**) after 20 min of vortexing. Scale bars: 50 µm.

The size and morphology of the crystals influence mixing dynamics and diffusion time scales. Iq-mEmerald crystal slurries can be reproducibly prepared in various morphologies, including needles, rectangular nuggets, and cubes, with various sizes, making them an ideal standard sample for applications such as nozzle design and jetting tests.

For iq-mEmerald, a lower $(NH_4)_2SO_4$ concentration leads to the formation of needle crystals. The process is independent of adding seedstock. The transition of the crystal morphology happens between the addition of 2 M and 2.5 M $(NH_4)_2SO_4$ crystallization solutions (Figure 4).

The size of cubic crystals can be adjusted by varying the concentration of $(NH_4)_2SO_4$ without altering the protein concentration. The crystals mature overnight and remain stable at ambient temperature over extended time periods.

Fluorescence quenching in iq-mEmerald crystals was observed upon mixing them using a mix-and-inject nozzle [63]. To quench the fluorescence of iq-mEmerald microcrystals, a 20 mM $CuCl_2$ (in 50 mM Tris, pH 8.0) solution can be used. The quenching can be used to study different nozzles or the diffusion time in various media using a high-speed camera. For the analysis of a mixing HVE nozzle, iq-mEmerald needle crystals were embedded in LCP and the injection was stable with a clear fluorescence signal from crystals (Figure S1E). A 20 mM $CuCl_2$ solution, also embedded in LCP, was used to assess successful mixing using the mixer at different flow rates/mixing times.

For the injection of the sample using a DFFN, 15% (*v/v*) cell pellet in the crystallization solution was used and a stable jet was produced (Figure S1D), leading to a reasonable hit-rate (not published). The flexibility of crystallization morphologies and the ability of fluorescence quenching in crystals make iq-mEmerald a potentially useful new standard sample for various applications.

Figure 4. Iq-mEmerald crystal morphology is dependent on the $(NH_4)_2SO_4$ concentration. The depicted values above the respective images relate to the respective precipitant solutions, not the final $(NH_4)_2SO_4$ concentrations of the crystallization setups. Scale bars: 10 μm.

3.4. PYP

Photoactive Yellow protein (PYP) from *Ectothiorhodospira halophila* is a well-studied photoactive protein used in many early time-resolved experiments using pump-probe approaches [66,73,112,113]. PYP is dynamic, showing a reversible photocycle in crystals, and does not need to be produced or handled in the dark. This makes it an ideal standard sample for SFX experiments.

Our protocol is on the basis of Schmidt et al. (2019) and Kim et al. (2013) [105,106] but with modifications, leading to higher quantities (over 50 mg/L from 1 L cell culture) as reported before [114,115]. We also adapted the production of the chromophore (*p*CA) from Thomson et al. (2019) [116] and Kim et al. (2013) [106] to streamline the production and purification time to two days (excluding expression).

For crystallization, the use of seedstock leads to the rapid production of large quantities of highly homogeneous microcrystals, usually within a few hours.

3.4.1. Upscaling of Protein Production

As huge amounts of protein are needed for a conclusive time-resolved experiment with several time-points, a robust, high-efficiency expression and purification protocol is important for standard samples. As the purification of PYP is rather cumbersome, it was optimized where possible.

The growth curves of Rosetta(DE3)-PYP and BL21(DE3)-PYP revealed that after induction with IPTG, the growth of BL21(DE3)-PYP was slowed down, leading to lower cell volumes in comparison with Rosetta(DE3)-PYP (Figure 5A). Higher gene expression values in Rosetta(DE3)-PYP cells were also visible on SDS-PAGE analysis (Figure 5B). Hence, Rosetta(DE3)-PYP was used as the expression host.

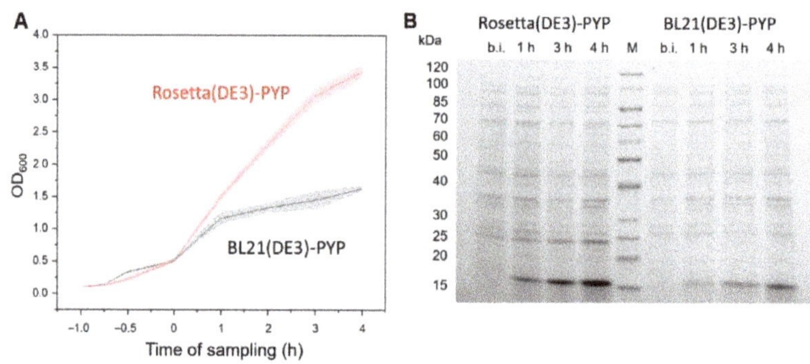

Figure 5. Recombinant expression of *pyp* in different *E. coli* expression strains. (**A**) Growth curves of Rosetta(DE3)-PYP and BL21(DE3)-PYP. For each strain, two single colonies were used to inoculate a pre-culture each. From each pre-culture, three main cultures were inoculated, respectively, with a start OD_{600} of 0.1. The OD_{600} was measured 60 min and 30 min before the induction, as well as 1 h, 2 h, 3 h and 4 h after induction with 1 mM IPTG. The lines depict the OD_{600} averages for each expression strain, whereas the standard deviations are given in the shaded areas. (**B**) SDS-PAGE comparison of recombinant PYP (~14 kDa) expression. M: Marker, 1 h, etc., depict the time of sampling after induction with 1 mM IPTG, b.i.: before induction with IPTG. The raw SDS-PAGE image is provided in Figure S2.

3.4.2. Seedstock Preparation and Crystallization

To make the crystallization quick, efficient and robust, seedstock was produced, as our initial crystallization trials led to inhomogeneous crystal slurries.

A large volume of seedstock was produced from an inhomogeneous batch crystallization setup. Several rounds of vortexing were needed to achieve a suitable seedstock (Figure 6).

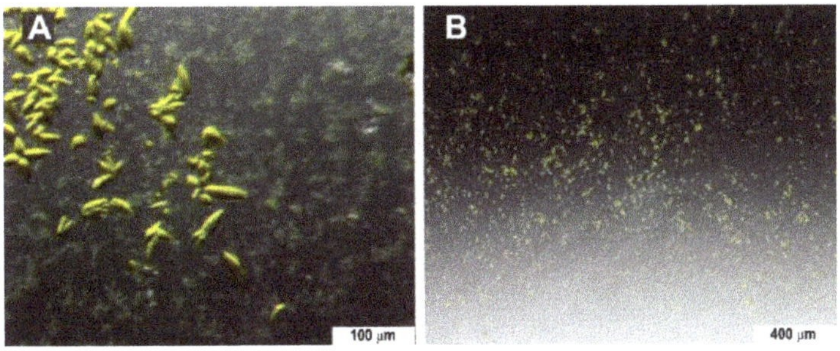

Figure 6. PYP crystals. (**A**) Inhomogeneous large crystals were obtained and used as a starting suspension for the preparation of seedstock. (**B**) Homogeneous PYP microcrystals obtained using seedstock.

Using different seedstock volumes, we could modulate the size of the crystals and crystals were ready in 1 h after crystallization was setup, allowing modifications of crystal size even during an ongoing experiment.

The crystals float on top of the solution due to the higher density of the crystallization solution, making it difficult to accurately estimate the crystal density. To generate a stable liquid jet, the sample volume was adjusted to 1.5 times compared to the total volume used in crystallization (Figure S1F).

The optimized expression and purification protocol, as well as the crystallization using seedstock, led to a reliable and reproducible production of PYP microcrystals in large amounts, making them a useful standard sample for pump-probe experiments.

4. Discussion

In this review, we present detailed and reproducible protocols for the preparation and crystallization of lysozyme, myoglobin, iq-mEmerald, and PYP, specifically adapted for SFX. These standard samples were chosen for their straightforward crystallization properties, well-characterized structural data, and relevance to a wide range of SFX applications, including time-resolved studies and instrument commissioning.

Each protein system required specific modifications to optimize crystal quality and to make suitable sample delivery methods. Lysozyme, as a widely used benchmark sample, was crystallized under controlled conditions to ensure consistent size and density. The size of the crystals can be easily modulated by using different temperatures, without the need to change the concentration of precipitant solutions or Lysozyme. Myoglobin crystallization was refined by precise adjustment of precipitant concentrations and, where necessary, deoxygenation steps to support studies of redox-state and ligand binding. The iq-mEmerald system, with its intrinsic fluorescence, provided real-time monitoring of sample injections and diffusion in mixing nozzles as well as the ability to easily produce crystals with different sizes and morphologies. Lastly, we provided a PYP preparation protocol of improved yield and shortened purification and crystallization times. The crystallization of PYP required using a seedstock, which significantly improved crystal homogeneity and growth reproducibility.

All four standard samples, lysozyme, myoglobin, iq-mEmerald, and PYP, exhibit stable crystal suspensions for a minimum of six months when stored at room temperature. To ensure optimal sample quality prior to use, an additional filtration step is performed immediately before injection to remove any large particles or aggregates that may have formed during storage. This approach maintains consistent crystal size distribution and preserves sample homogeneity, critical for reliable serial crystallography experiments. This long-term also benefits their use as convenient and reliable standard samples for serial crystallography experiments.

Each of these protein crystal samples has been validated through data collection at the SPB/SFX instrument at the European XFEL. Under standard experimental conditions, lysozyme, myoglobin, and iq-mEmerald crystals diffracted to approximately 1.7 Å, 1.6 Å, and 1.8 Å resolution at 9.3 keV, respectively, while PYP crystals reached 1.3 Å at 11.56 keV. These high-quality diffraction results confirm the suitability of these proteins as standard samples for serial crystallography. It is noted that even higher resolutions may be attainable, as the current limits were set by the used photon energy and detector distance rather than intrinsic crystal quality. Together, these outcomes demonstrate both the practical reliability and benchmarking value of the selected standards for SFX applications at XFEL facilities.

The standardized methods presented here not only facilitate accurate and reproducible data collection but also support the broader adoption of SFX by lowering technical barriers for new users and laboratories. By providing clear, stepwise protocols, this work contributes to the harmonization of sample preparation practices across the SFX community, enabling more consistent benchmarking and comparison of experimental results.

While our focus remains on lysozyme, myoglobin, iq-mEmerald, and PYP, we also highlight other proteins and biological particles, such as thermolysin, glucose isomerase, proteinase K, trypsin, other GFP derivatives, thaumatin, and granulovirus that serve as alternative or complementary standards for SFX. These additional samples offer unique properties and extend the versatility of SFX for a wider range of structural biology challenges.

5. Conclusions

The protocols presented in this review provide a reliable framework for preparing high-quality standard protein crystals suitable for SFX experiments. The standardized methods contribute to the advancement of structural biology by enabling accurate and reproducible data collection and facilitating proof-of-concept studies as well as instrument commissioning. By streamlining such applications, these processes accelerate the generation of biologically meaningful structural insights and contribute to a deeper understanding of biomolecular structure and dynamics.

Supplementary Materials: The following supporting information can be downloaded at https://www.mdpi.com/article/10.3390/biom15111488/s1.mdpi.com/article/doi/s1. Figure S1: Sample injection images; Figure S2: The unprocessed SDS-PAGE image used in Figure 5. References [117,118] are cited in the Supplementary Materials.

Author Contributions: Conceptualization, C.S. and H.H.; investigation, C.S. and H.H.; resources, K.L. and J.S.; writing—original draft preparation, H.H. and C.S.; writing—review and editing, H.H., C.S., K.L. and J.S. All authors have read and agreed to the published version of the manuscript.

Funding: This research received no external funding.

Institutional Review Board Statement: Not applicable.

Informed Consent Statement: Not applicable.

Acknowledgments: We gratefully acknowledge all members of the SEC group for their excellent teamwork, support, and insightful discussions throughout this work. We also thank the SPB/SFX team for their valued collaboration and contributions to this study. Special appreciation is owed to Raphael de Wijn (SPB/SFX) for generously sharing data analysis results that greatly enhanced our research.

Conflicts of Interest: All authors were employed by the company European XFEL GmbH. The remaining authors declare that the research was conducted in the absence of any commercial or financial relationships that could be construed as a potential conflict of interest.

References

1. Chapman, H.N.; Fromme, P.; Barty, A.; White, T.A.; Kirian, R.A.; Aquila, A.; Hunter, M.S.; Schulz, J.; Deponte, D.P.; Weierstall, U.; et al. Femtosecond X-Ray Protein Nanocrystallography. *Nature* **2011**, *470*, 73–77. [CrossRef]
2. Neutze, R.; Wouts, R.; Van Der Spoel, D.; Weckert, E.; Hajdu, J. Potential for Biomolecular Imaging with Femtosecond X-Ray Pulses. *Nature* **2000**, *406*, 752–757. [CrossRef]
3. Barends, T.R.M.; Stauch, B.; Cherezov, V.; Schlichting, I. Serial Femtosecond Crystallography. *Nat. Rev. Methods Prim.* **2022**, *2*, 59. [CrossRef]
4. Orville, A.M. Recent Results in Time Resolved Serial Femtosecond Crystallography at XFELs. *Curr. Opin. Struct. Biol.* **2020**, *65*, 193–208. [CrossRef]
5. Kupitz, C.; Sierra, R.G. Preventing Bio-Bloopers and XFEL Follies: Best Practices from Your Friendly Instrument Staff. *Crystals* **2020**, *10*, 251. [CrossRef]
6. Botha, S.; Fromme, P. Review of Serial Femtosecond Crystallography Including the COVID-19 Pandemic Impact and Future Outlook. *Structure* **2023**, *31*, 1306–1319. [CrossRef] [PubMed]
7. Lee, D.B.; Kim, J.M.; Seok, J.H.; Lee, J.H.; Jo, J.D.; Mun, J.Y.; Conrad, C.; Coe, J.; Nelson, G.; Hogue, B.; et al. Supersaturation-Controlled Microcrystallization and Visualization Analysis for Serial Femtosecond Crystallography. *Sci. Rep.* **2018**, *8*, 2541. [CrossRef] [PubMed]
8. Fromme, R.; Ishchenko, A.; Metz, M.; Chowdhury, S.R.; Basu, S.; Boutet, S.; Fromme, P.; White, T.A.; Barty, A.; Spence, J.C.H.; et al. Serial Femtosecond Crystallography of Soluble Proteins in Lipidic Cubic Phase. *IUCrJ* **2015**, *2*, 545–551. [CrossRef]
9. Stellato, F.; Oberthür, D.; Liang, M.; Bean, R.; Gati, C.; Yefanov, O.; Barty, A.; Burkhardt, A.; Fischer, P.; Galli, L.; et al. Room-Temperature Macromolecular Serial Crystallography Using Synchrotron Radiation. *IUCrJ* **2014**, *1*, 204–212. [CrossRef]
10. Perrett, S.; Fadini, A.; Hutchison, C.D.M.; Bhattacharya, S.; Morrison, C.; Turkot, O.; Jakobsen, M.B.; Großler, M.; Licon-Salaiz, J.; Griese, F.; et al. Kilohertz Droplet-on-Demand Serial Femtosecond Crystallography at the European XFEL Station FX. *Struct. Dyn.* **2024**, *11*, 024310.
11. DePonte, D.P.; Weierstall, U.; Schmidt, K.; Warner, J.; Starodub, D.; Spence, J.C.H.; Doak, R.B. Gas Dynamic Virtual Nozzle for Generation of Microscopic Droplet Streams. *J. Phys. D Appl. Phys.* **2008**, *41*, 195505. [CrossRef]
12. Tolstikova, A.; Levantino, M.; Yefanov, O.; Hennicke, V.; Fischer, P.; Meyer, J.; Mozzanica, A.; Redford, S.; Crosas, E.; Opara, N.L.; et al. 1 KHz Fixed-Target Serial Crystallography Using a Multilayer Monochromator and an Integrating Pixel Detector. *IUCrJ* **2019**, *6*, 927–937. [CrossRef]
13. Sierra, R.G.; Gati, C.; Laksmono, H.; Dao, E.H.; Gul, S.; Fuller, F.; Kern, J.; Chatterjee, R.; Ibrahim, M.; Brewster, A.S.; et al. Concentric-Flow Electrokinetic Injector Enables Serial Crystallography of Ribosome and Photosystem II. *Nat. Methods* **2015**, *13*, 59–62. [CrossRef] [PubMed]
14. Sierra, R.G.; Laksmono, H.; Kern, J.; Tran, R.; Hattne, J.; Alonso-Mori, R.; Lassalle-Kaiser, B.; Glöckner, C.; Hellmich, J.; Schafer, D.W.; et al. Nanoflow Electrospinning Serial Femtosecond Crystallography. *Acta Crystallogr. Sect. D Biol. Crystallogr.* **2012**, *68*, 1584–1587. [CrossRef] [PubMed]
15. Lieske, J.; Cerv, M.; Kreida, S.; Komadina, D.; Fischer, J.; Barthelmess, M.; Fischer, P.; Pakendorf, T.; Yefanov, O.; Mariani, V.; et al. On-Chip Crystallization for Serial Crystallography Experiments and on-Chip Ligand-Binding Studies. *IUCrJ* **2019**, *6*, 714–728. [CrossRef] [PubMed]
16. Roessler, C.G.; Agarwal, R.; Allaire, M.; Alonso-Mori, R.; Andi, B.; Bachega, J.F.R.; Bommer, M.; Brewster, A.S.; Browne, M.C.; Chatterjee, R.; et al. Acoustic Injectors for Drop-On-Demand Serial Femtosecond Crystallography. *Structure* **2016**, *24*, 631–640. [CrossRef]
17. Park, J.; Park, S.; Kim, J.; Park, G.; Cho, Y.; Nam, K.H. Polyacrylamide Injection Matrix for Serial Femtosecond Crystallography. *Sci. Rep.* **2019**, *9*, 2525. [CrossRef]
18. Naitow, H.; Matsuura, Y.; Tono, K.; Joti, Y.; Kameshima, T.; Hatsui, T.; Yabashi, M.; Tanaka, R.; Tanaka, T.; Sugahara, M.; et al. Protein-Ligand Complex Structure from Serial Femtosecond Crystallography Using Soaked Thermolysin Microcrystals and Comparison with Structures from Synchrotron Radiation. *Acta Crystallogr. Sect. D Struct. Biol.* **2017**, *73*, 702–709. [CrossRef]
19. Matthews, B.W.; Weaver, L.H.; Kester, W.R. The Conformation of Thermolysin. *J. Biol. Chem.* **1974**, *249*, 8030–8044. [CrossRef]
20. Dahlquist, F.W.; Long, J.W.; Bigbee, W.L. Role of Calcium in the Thermal Stability of Thermolysin. *Biochemistry* **1976**, *15*, 1103–1111. [CrossRef]
21. Roche, R.S.; Voordouw, G.; Matthews, B.W. The Structural and Functional Roles of Metal Ions in Thermolysin. *Crit. Rev. Biochem. Mol. Biol.* **1978**, *5*, 1–23. [CrossRef]
22. Leite, J.P.; Gales, L. Alzheimer's Aβ 1-40 Peptide Degradation by Thermolysin: Evidence of Inhibition by a C-Terminal Aβ Product. *FEBS Lett.* **2019**, *593*, 128–137. [CrossRef] [PubMed]
23. Kovalevsky, A.Y.; Hanson, L.; Fisher, S.Z.; Mustyakimov, M.; Mason, S.A.; Trevor Forsyth, V.; Blakeley, M.P.; Keen, D.A.; Wagner, T.; Carrell, H.L.; et al. Metal Ion Roles and the Movement of Hydrogen during Reaction Catalyzed by D-Xylose Isomerase: A Joint X-Ray and Neutron Diffraction Study. *Structure* **2010**, *18*, 688–699. [CrossRef]

24. Callens, M.; Kersters-Hilderson, H.; Van Opstal, O.; De Bruyne, C.K. Catalytic Properties of D-Xylose Isomerase from Streptomyces Violaceoruber. *Enzym. Microb. Technol.* **1986**, *8*, 696–700. [CrossRef]
25. Danno, G. ichi Studies on D-Glucose-Isomerizing Enzyme of Bacillus Coagulans, Strain HN-68: Part III. Induced Formation of D-Glucose-Isomerizing Enzyme by D-Glucose-Grown Cells of Bacillus Coagulans, Strain HN-68. *Agric. Biol. Chem.* **1970**, *34*, 1658–1667. [CrossRef]
26. Nam, K.H. Stable Sample Delivery in Viscous Media via a Capillary for Serial Crystallography. *J. Appl. Crystallogr.* **2020**, *53*, 45–50. [CrossRef]
27. Nam, K.H. Shortening Injection Matrix for Serial Crystallography. *Sci. Rep.* **2020**, *10*, 107. [CrossRef]
28. Nam, K.H. Polysaccharide-Based Injection Matrix for Serial Crystallography. *Int. J. Mol. Sci.* **2020**, *21*, 3332. [CrossRef]
29. Nam, K.H. Beef Tallow Injection Matrix for Serial Crystallography. *Sci. Rep.* **2022**, *12*, 694. [CrossRef]
30. Sugahara, M.; Mizohata, E.; Nango, E.; Suzuki, M.; Tanaka, T.; Masuda, T.; Tanaka, R.; Shimamura, T.; Tanaka, Y.; Suno, C.; et al. Grease Matrix as a Versatile Carrier of Proteins for Serial Crystallography. *Nat. Methods* **2015**, *12*, 61–63. [CrossRef]
31. Lee, D.; Baek, S.; Park, J.; Lee, K.; Kim, J.; Lee, S.J.; Chung, W.K.; Lee, J.L.; Cho, Y.; Nam, K.H. Nylon Mesh-Based Sample Holder for Fixed-Target Serial Femtosecond Crystallography. *Sci. Rep.* **2019**, *9*, 6971. [CrossRef]
32. Kim, Y.; Nam, K.H. Fixed-Target Pink-Beam Serial Synchrotron Crystallography at Pohang Light Source II. *Crystals* **2023**, *13*, 1544. [CrossRef]
33. Mehrabi, P.; Schulz, E.C.; Agthe, M.; Horrell, S.; Bourenkov, G.; von Stetten, D.; Leimkohl, J.P.; Schikora, H.; Schneider, T.R.; Pearson, A.R.; et al. Liquid Application Method for Time-Resolved Analyses by Serial Synchrotron Crystallography. *Nat. Methods* **2019**, *16*, 979–982. [CrossRef]
34. Mehrabi, P.; Sung, S.; von Stetten, D.; Prester, A.; Hatton, C.E.; Kleine-Döpke, S.; Berkes, A.; Gore, G.; Leimkohl, J.P.; Schikora, H.; et al. Millisecond Cryo-Trapping by the Spitrobot Crystal Plunger Simplifies Time-Resolved Crystallography. *Nat. Commun.* **2023**, *14*, 2365. [CrossRef]
35. Norton-Baker, B.; Mehrabi, P.; Boger, J.; Schönherr, R.; Von Stetten, D.; Schikora, H.; Kwok, A.O.; Martin, R.W.; Miller, R.J.D.; Redecked, L.; et al. A Simple Vapor-Diffusion Method Enables Protein Crystallization inside the HARE Serial Crystallography Chip. *Acta Crystallogr. Sect. D Struct. Biol.* **2021**, *77*, 820–834. [CrossRef] [PubMed]
36. Martin-Garcia, J.M.; Zhu, L.; Mendez, D.; Lee, M.Y.; Chun, E.; Li, C.; Hu, H.; Subramanian, G.; Kissick, D.; Ogata, C.; et al. High-Viscosity Injector-Based Pink-Beam Serial Crystallography of Microcrystals at a Synchrotron Radiation Source. *IUCrJ* **2019**, *6*, 412–425. [CrossRef]
37. Meents, A.; Wiedorn, M.O.; Srajer, V.; Henning, R.; Sarrou, I.; Bergtholdt, J.; Barthelmess, M.; Reinke, P.Y.A.; Dierksmeyer, D.; Tolstikova, A.; et al. Pink-Beam Serial Crystallography. *Nat. Commun.* **2017**, *8*, 1281. [CrossRef]
38. Botha, S.; Baitan, D.; Jungnickel, K.E.J.; Oberthür, D.; Schmidt, C.; Stern, S.; Wiedorn, M.O.; Perbandt, M.; Chapman, H.N.; Betzel, C. De Novo Protein Structure Determination by Heavy-Atom Soaking in Lipidic Cubic Phase and SIRAS Phasing Using Serial Synchrotron Crystallography. *IUCrJ* **2018**, *5*, 524–530. [CrossRef]
39. Martin-Garcia, J.M.; Conrad, C.E.; Nelson, G.; Stander, N.; Zatsepin, N.A.; Zook, J.; Zhu, L.; Geiger, J.; Chun, E.; Kissick, D.; et al. Serial Millisecond Crystallography of Membrane and Soluble Protein Microcrystals Using Synchrotron Radiation. *IUCrJ* **2017**, *4*, 439–454. [CrossRef] [PubMed]
40. Orlans, J.; Rose, S.L.; Ferguson, G.; Oscarsson, M.; Homs Puron, A.; Beteva, A.; Debionne, S.; Theveneau, P.; Coquelle, N.; Kieffer, J.; et al. Advancing Macromolecular Structure Determination with Microsecond X-Ray Pulses at a 4th Generation Synchrotron. *Commun. Chem.* **2025**, *8*, 6. [CrossRef]
41. Zhao, F.Z.; Sun, B.; Yu, L.; Xiao, Q.J.; Wang, Z.J.; Chen, L.L.; Liang, H.; Wang, Q.S.; He, J.H.; Yin, D.C. A Novel Sample Delivery System Based on Circular Motion for: In Situ Serial Synchrotron Crystallography. *Lab. Chip* **2020**, *20*, 3888–3898. [CrossRef] [PubMed]
42. Lee, D.; Park, S.; Lee, K.; Kim, J.; Park, G.; Nam, K.H.; Baek, S.; Chung, W.K.; Lee, J.L.; Cho, Y.; et al. Application of a High-Throughput Microcrystal Delivery System to Serial Femtosecond Crystallography. *J. Appl. Crystallogr.* **2020**, *53*, 477–485. [CrossRef] [PubMed]
43. Okumura, H.; Sakai, N.; Murakami, H.; Mizuno, N.; Nakamura, Y.; Ueno, G.; Masunaga, T.; Kawamura, T.; Baba, S.; Hasegawa, K.; et al. In Situ Crystal Data-Collection and Ligand-Screening System at SPring-8. *Acta Crystallogr. Sect. F Struct. Biol. Commun.* **2022**, *78*, 241–251. [CrossRef]
44. Yin, X.; Scalia, A.; Leroy, L.; Cuttitta, C.M.; Polizzo, G.M.; Ericson, D.L.; Roessler, C.G.; Campos, O.; Ma, M.Y.; Agarwal, R.; et al. Hitting the Target: Fragment Screening with Acoustic in Situ Co-Crystallization of Proteins plus Fragment Libraries on Pin-Mounted Data-Collection Micromeshes. *Acta Crystallogr. Sect. D Biol. Crystallogr.* **2014**, *70*, 1177–1189. [CrossRef]
45. Stubbs, J.; Hornsey, T.; Hanrahan, N.; Esteban, L.B.; Bolton, R.; Malý, M.; Basu, S.; Orlans, J.; de Sanctis, D.; Shim, J.U.; et al. Droplet Microfluidics for Time-Resolved Serial Crystallography. *IUCrJ* **2024**, *11*, 237–248. [CrossRef]

46. Gao, Y.; Xu, W.; Shi, W.; Soares, A.; Jakoncic, J.; Myers, S.; Martins, B.; Skinner, J.; Liu, Q.; Bernstein, H.; et al. High-Speed Raster-Scanning Synchrotron Serial Microcrystallography with a High-Precision Piezo-Scanner. *J. Synchrotron Radiat.* **2018**, *25*, 1362–1370. [CrossRef]
47. Kendrew, J.C.; Bodo, G.; Dintzis, H.M.; Parrish, R.G.; Wyckoff, H.; Phillips, D.C. A Three-Dimensional Model of the Myoglobin Molecule Obtained by x-Ray Analysis. *Nature* **1958**, *181*, 662–666. [CrossRef] [PubMed]
48. Barends, T.R.M.; Foucar, L.; Ardevol, A.; Nass, K.; Aquila, A.; Botha, S.; Doak, R.B.; Falahati, K.; Hartmann, E.; Hilpert, M.; et al. Direct Observation of Ultrafast Collective Motions in CO Myoglobin upon Ligand Dissociation. *Science* **2015**, *350*, 445–450. [CrossRef] [PubMed]
49. Levantino, M.; Schirò, G.; Lemke, H.T.; Cottone, G.; Glownia, J.M.; Zhu, D.; Chollet, M.; Ihee, H.; Cupane, A.; Cammarata, M. Ultrafast Myoglobin Structural Dynamics Observed with an X-Ray Free-Electron Laser. *Nat. Commun.* **2015**, *6*, 6772. [CrossRef]
50. Owen, R.L.; Axford, D.; Sherrell, D.A.; Kuo, A.; Ernst, O.P.; Schulz, E.C.; Miller, R.J.D.; Mueller-Werkmeister, H.M. Low-Dose Fixed-Target Serial Synchrotron Crystallography. *Acta Crystallogr. Sect. D Struct. Biol.* **2017**, *73*, 373–378. [CrossRef]
51. Oghbaey, S.; Sarracini, A.; Ginn, H.M.; Pare-Labrosse, O.; Kuo, A.; Marx, A.; Epp, S.W.; Sherrell, D.A.; Eger, B.T.; Zhong, Y.; et al. Fixed Target Combined with Spectral Mapping: Approaching 100% Hit Rates for Serial Crystallography. *Acta Crystallogr. Sect. D Struct. Biol.* **2016**, *72*, 944–955. [CrossRef]
52. Mehrabi, P.; Bücker, R.; Bourenkov, G.; Ginn, H.M.; von Stetten, D.; Müller-Werkmeister, H.M.; Kuo, A.; Morizumi, T.; Eger, B.T.; Ou, W.L.; et al. Serial Femtosecond and Serial Synchrotron Crystallography Can Yield Data of Equivalent Quality: A Systematic Comparison. *Sci. Adv.* **2021**, *7*, eabf1380. [CrossRef]
53. Ebrahim, A.; Moreno-Chicano, T.; Appleby, M.V.; Chaplin, A.K.; Beale, J.H.; Sherrell, D.A.; Duyvesteyn, H.M.E.; Owada, S.; Tono, K.; Sugimoto, H.; et al. Dose-Resolved Serial Synchrotron and XFEL Structures of Radiation-Sensitive Metalloproteins. *IUCrJ* **2019**, *6*, 543–551. [CrossRef]
54. Remington, S.J. Green Fluorescent Protein: A Perspective. *Protein Sci.* **2011**, *20*, 1509–1519. [CrossRef]
55. Lukyanov, K.A.; Chudakov, D.M.; Lukyanov, S.; Verkhusha, V.V. Photoactivatable Fluorescent Proteins. *Nat. Rev. Mol. Cell Biol.* **2005**, *6*, 885–890. [CrossRef] [PubMed]
56. Patterson, G.H.; Lippincott-Schwartz, J. A Photoactivatable GFP for Selective Photolabeling of Proteins and Cells. *Science* **2002**, *297*, 1873–1877. [CrossRef]
57. Frommer, W.B.; Davidson, M.W.; Campbell, R.E. Genetically Encoded Biosensors Based on Engineered Fluorescent Proteins. *Chem. Soc. Rev.* **2009**, *38*, 2833–2841. [CrossRef] [PubMed]
58. Coquelle, N.; Sliwa, M.; Woodhouse, J.; Schirò, G.; Adam, V.; Aquila, A.; Barends, T.R.M.; Boutet, S.; Byrdin, M.; Carbajo, S.; et al. Chromophore Twisting in the Excited State of a Photoswitchable Fluorescent Protein Captured by Time-Resolved Serial Femtosecond Crystallography. *Nat. Chem.* **2018**, *10*, 31–37. [CrossRef] [PubMed]
59. Woodhouse, J.; Nass Kovacs, G.; Coquelle, N.; Uriarte, L.M.; Adam, V.; Barends, T.R.M.; Byrdin, M.; de la Mora, E.; Bruce Doak, R.; Feliks, M.; et al. Photoswitching Mechanism of a Fluorescent Protein Revealed by Time-Resolved Crystallography and Transient Absorption Spectroscopy. *Nat. Commun.* **2020**, *11*, 741. [CrossRef]
60. Adam, V.; Hadjidemetriou, K.; Jensen, N.; Shoeman, R.L.; Woodhouse, J.; Aquila, A.; Banneville, A.S.; Barends, T.R.M.; Bezchastnov, V.; Boutet, S.; et al. Rational Control of Off-State Heterogeneity in a Photoswitchable Fluorescent Protein Provides Switching Contrast Enhancement**. *ChemPhysChem* **2022**, *23*, e202200192. [CrossRef]
61. Beale, J.H.; Bolton, R.; Marshall, S.A.; Beale, E.V.; Carr, S.B.; Ebrahim, A.; Moreno-Chicano, T.; Hough, M.A.; Worrall, J.A.R.; Tews, I.; et al. Successful Sample Preparation for Serial Crystallography Experiments. *J. Appl. Crystallogr.* **2019**, *52*, 1385–1396. [CrossRef]
62. Yu, X.; Strub, M.P.; Barnard, T.J.; Noinaj, N.; Piszczek, G.; Buchanan, S.K.; Taraska, J.W. An Engineered Palette of Metal Ion Quenchable Fluorescent Proteins. *PLoS ONE* **2014**, *9*, e95808. [CrossRef]
63. Vakili, M.; Han, H.; Schmidt, C.; Wrona, A.; Kloos, M.; De Diego, I.; Dörner, K.; Geng, T.; Kim, C.; Koua, F.H.M.; et al. Mix-and-Extrude: High-Viscosity Sample Injection towards Time-Resolved Protein Crystallography. *J. Appl. Crystallogr.* **2023**, *56*, 1038–1045. [CrossRef]
64. Hutchison, C.D.M.; van Thor, J.J. Populations and Coherence in Femtosecond Time Resolved X-Ray Crystallography of the Photoactive Yellow Protein. *Int. Rev. Phys. Chem.* **2017**, *36*, 117–143. [CrossRef]
65. Konold, P.E.; Arik, E.; Weißenborn, J.; Arents, J.C.; Hellingwerf, K.J.; van Stokkum, I.H.M.; Kennis, J.T.M.; Groot, M.L. Confinement in Crystal Lattice Alters Entire Photocycle Pathway of the Photoactive Yellow Protein. *Nat. Commun.* **2020**, *11*, 4248. [CrossRef]
66. Pandey, S.; Bean, R.; Sato, T.; Poudyal, I.; Bielecki, J.; Cruz Villarreal, J.; Yefanov, O.; Mariani, V.; White, T.A.; Kupitz, C.; et al. Time-Resolved Serial Femtosecond Crystallography at the European XFEL. *Nat. Methods* **2020**, *17*, 73–78. [CrossRef] [PubMed]
67. Schmidt, M.; Pande, K.; Basu, S.; Tenboer, J. Room Temperature Structures beyond 1.5 Å by Serial Femtosecond Crystallography. *Struct. Dyn.* **2015**, *2*, 041708. [CrossRef]

68. Schotte, F.; Cho, H.S.; Kaila, V.R.I.; Kamikubo, H.; Dashdorj, N.; Henry, E.R.; Graber, T.J.; Henning, R.; Wulff, M.; Hummer, G. et al. Watching a Signaling Protein Function in Real Time via 100-Ps Time-Resolved Laue Crystallography. *Proc. Natl. Acad. Sci USA* **2012**, *109*, 19256–19261. [CrossRef]
69. Ihee, H.; Rajagopal, S.; Srajer, V.; Pahl, R.; Anderson, S.; Schmidt, M.; Schotte, F.; Anfinrud, P.A.; Wulff, M.; Moffat, K. Visualizing Reaction Pathways in Photoactive Yellow Protein from Nanoseconds to Seconds. *Proc. Natl. Acad. Sci. USA* **2005**, *102*, 7145–7150 [CrossRef] [PubMed]
70. Liu, J.; Yabushita, A.; Taniguchi, S.; Chosrowjan, H.; Imamoto, Y.; Sueda, K.; Miyanaga, N.; Kobayashi, T. Ultrafast Time-Resolved Pump-Probe Spectroscopy of PYP by a Sub-8 Fs Pulse Laser at 400 Nm. *J. Phys. Chem. B* **2013**, *117*, 4818–4826. [CrossRef]
71. Creelman, M.; Kumauchi, M.; Hoff, W.D.; Mathies, R.A. Chromophore Dynamics in the PYP Photocycle from Femtosecond Stimulated Raman Spectroscopy. *J. Phys. Chem. B* **2014**, *118*, 659–667. [CrossRef]
72. Liu, Z.; Gu, K.; Shelby, M.; Roy, D.; Muniyappan, S.; Schmidt, M.; Narayanasamy, S.R.; Coleman, M.; Frank, M.; Kuhl, T.L. In Situ Counter-Diffusion Crystallization and Long-Term Crystal Preservation in Microfluidic Fixed Targets for Serial Crystallography. *J Appl. Crystallogr.* **2024**, *57*, 1539–1550. [CrossRef]
73. Kim, T.W.; Lee, J.H.; Choi, J.; Kim, K.H.; Van Wilderen, L.J.; Guerin, L.; Kim, Y.; Jung, Y.O.; Yang, C.; Kim, J.; et al. Protein Structural Dynamics of Photoactive Yellow Protein in Solution Revealed by Pump-Probe X-Ray Solution Scattering. *J. Am. Chem Soc.* **2012**, *134*, 3145–3153. [CrossRef]
74. Kim, J.G.; Kim, T.W.; Kim, J.; Ihee, H. Protein Structural Dynamics Revealed by Time-Resolved X-Ray Solution Scattering. *Acc Chem. Res.* **2015**, *48*, 2200–2208. [CrossRef] [PubMed]
75. Foos, N.; Seuring, C.; Schubert, R.; Burkhardt, A.; Svensson, O.; Meents, A.; Chapman, H.N.; Nanao, M.H. X-Ray and UV Radiation-Damage-Induced Phasing Using Synchrotron Serial Crystallography. *Acta Crystallogr. Sect. D Struct. Biol.* **2018**, *74* 366–378. [CrossRef]
76. Gilbile, D.; Shelby, M.L.; Lyubimov, A.Y.; Wierman, J.L.; Monteiro, D.C.F.; Cohen, A.E.; Russi, S.; Coleman, M.A.; Frank, M.; Kuhl T.L. Plug-and-Play Polymer Microfluidic Chips for Hydrated, Room-Temperature Fixed-Target Serial Crystallography. *Lab. Chip* **2021**, *21*, 4831–4845. [CrossRef]
77. Nass, K.; Meinhart, A.; Barends, T.R.M.; Foucar, L.; Gorel, A.; Aquila, A.; Botha, S.; Doak, R.B.; Koglin, J.; Liang, M.; et al. Protein Structure Determination by Single-Wavelength Anomalous Diffraction Phasing of {X}-Ray Free-Electron Laser Data. *IUCrJ* **2016** *3*, 180–191. [CrossRef]
78. Nass, K.; Bacellar, C.; Cirelli, C.; Dworkowski, F.; Gevorkov, Y.; James, D.; Johnson, P.J.M.; Kekilli, D.; Knopp, G.; Martiel, I.; et al Pink-Beam Serial Femtosecond Crystallography for Accurate Structure-Factor Determination at an {X}-Ray Free-Electron Laser *IUCrJ* **2021**, *8*, 905–920. [CrossRef] [PubMed]
79. Williams, L.J.; Thompson, A.J.; Dijkstal, P.; Appleby, M.; Assmann, G.; Dworkowski, F.S.N.; Hiller, N.; Huang, C.-Y.; Mason, T. Perrett, S.; et al. Damage before Destruction? {X}-Ray-Induced Changes in Single-Pulse Serial Femtosecond Crystallography *IUCrJ* **2025**, *12*, 358–371. [CrossRef]
80. Liu, Z.; Gu, K.K.; Shelby, M.L.; Gilbile, D.; Lyubimov, A.Y.; Russi, S.; Cohen, A.E.; Narayanasamy, S.R.; Botha, S.; Kupitz, C.; et al A User-Friendly Plug-and-Play Cyclic Olefin Copolymer-Based Microfluidic Chip for Room-Temperature, Fixed-Target Serial Crystallography. *Acta Crystallogr. D Struct. Biol.* **2023**, *79*, 944–952. [CrossRef]
81. Jaho, S.; Sallaz-Damaz, Y.; Budayova-Spano, M. Microdialysis On-Chip Crystallization of Soluble and Membrane Proteins with the MicroCrys Platform and in Situ X-Ray Diffraction Case Studies. *CrystEngComm* **2023**, *25*, 5513–5523. [CrossRef]
82. Fan, J.; Jehle, J.A.; Wennmann, J.T. Population Structure of Cydia Pomonella Granulovirus Isolates Revealed by Quantitative Analysis of Genetic Variation. *Virus Evol.* **2021**, *7*, veaa073. [CrossRef]
83. Gati, C.; Oberthuer, D.; Yefanov, O.; Bunker, R.D.; Stellato, F.; Chiu, E.; Yeh, S.M.; Aquila, A.; Basu, S.; Bean, R.; et al. Atomic Structure of Granulin Determined from Native Nanocrystalline Granulovirus Using an X-Ray Free-Electron Laser. *Proc. Natl. Acad. Sci. USA* **2017**, *114*, 2247–2252. [CrossRef]
84. Awel, S.; Kirian, R.A.; Wiedorn, M.O.; Beyerlein, K.R.; Roth, N.; Horke, D.A.; Oberthür, D.; Knoska, J.; Mariani, V.; Morgan, A.; et al. Femtosecond X-Ray Diffraction from an Aerosolized Beam of Protein Nanocrystals. *J. Appl. Crystallogr.* **2018**, *51*, 133–139. [CrossRef]
85. Bücker, R.; Hogan-Lamarre, P.; Mehrabi, P.; Schulz, E.C.; Bultema, L.A.; Gevorkov, Y.; Brehm, W.; Yefanov, O.; Oberthür, D.; Kassier, G.H.; et al. Serial Protein Crystallography in an Electron Microscope. *Nat. Commun.* **2020**, *11*, 996. [CrossRef] [PubMed]
86. Boutet, S.; Lomb, L.; Williams, G.J.; Barends, T.R.M.; Aquila, A.; Doak, R.B.; Weierstall, U.; DePonte, D.P.; Steinbrener, J.; Shoeman, R.L.; et al. High-Resolution Protein Structure Determination by Serial Femtosecond Crystallography. *Science* **2012**, *337*, 362–364. [CrossRef] [PubMed]
87. Wiedorn, M.O.; Oberthür, D.; Bean, R.; Schubert, R.; Werner, N.; Abbey, B.; Aepfelbacher, M.; Adriano, L.; Allahgholi, A.; Al-Qudami, N.; et al. Megahertz Serial Crystallography. *Nat. Commun.* **2018**, *9*, 4025. [CrossRef]

88. Leonarski, F.; Nan, J.; Matej, Z.; Bertrand, Q.; Furrer, A.; Gorgisyan, I.; Bjelčić, M.; Kepa, M.; Glover, H.; Hinger, V.; et al. Kilohertz Serial Crystallography with the JUNGFRAU Detector at a Fourth-Generation Synchrotron Source. *IUCrJ* **2023**, *10*, 729–737. [CrossRef]
89. Martin-Garcia, J.M.; Botha, S.; Hu, H.; Jernigan, R.; Castellví, A.; Lisova, S.; Gil, F.; Calisto, B.; Crespo, I.; Roy-Chowdhury, S.; et al. Serial Macromolecular Crystallography at ALBA Synchrotron Light Source. *J. Synchrotron Radiat.* **2022**, *29*, 896–907. [CrossRef]
90. Monteiro, D.C.F.; Von Stetten, D.; Stohrer, C.; Sans, M.; Pearson, A.R.; Santoni, G.; Van Der Linden, P.; Trebbin, M. 3D-MiXD: 3D-Printed X-Ray-Compatible Microfluidic Devices for Rapid, Low-Consumption Serial Synchrotron Crystallography Data Collection in Flow. *IUCrJ* **2020**, *7*, 207–219. [CrossRef]
91. Galli, L.; Son, S.K.; Barends, T.R.M.; White, T.A.; Barty, A.; Botha, S.; Boutet, S.; Caleman, C.; Doak, R.B.; Nanao, M.H.; et al. Towards Phasing Using High X-Ray Intensity. *IUCrJ* **2015**, *2*, 627–634. [CrossRef]
92. Sugahara, M.; Nakane, T.; Masuda, T.; Suzuki, M.; Inoue, S.; Song, C.; Tanaka, R.; Nakatsu, T.; Mizohata, E.; Yumoto, F.; et al. Hydroxyethyl Cellulose Matrix Applied to Serial Crystallography. *Sci. Rep.* **2017**, *7*, 703. [CrossRef]
93. Beyerlein, K.R.; Dierksmeyer, D.; Mariani, V.; Kuhn, M.; Sarrou, I.; Ottaviano, A.; Awel, S.; Knoska, J.; Fuglerud, S.; Jönsson, O.; et al. Mix-and-Diffuse Serial Synchrotron Crystallography. *IUCrJ* **2017**, *4*, 769–777. [CrossRef]
94. Wranik, M.; Kepa, M.W.; Beale, E.V.; James, D.; Bertrand, Q.; Weinert, T.; Furrer, A.; Glover, H.; Gashi, D.; Carrillo, M.; et al. A Multi-Reservoir Extruder for Time-Resolved Serial Protein Crystallography and Compound Screening at X-Ray Free-Electron Lasers. *Nat. Commun.* **2023**, *14*, 7956. [CrossRef]
95. Wierman, J.L.; Paré-Labrosse, O.; Sarracini, A.; Besaw, J.E.; Cook, M.J.; Oghbaey, S.; Daoud, H.; Mehrabi, P.; Kriksunov, I.; Kuo, A.; et al. Fixed-Target Serial Oscillation Crystallography at Room Temperature. *IUCrJ* **2019**, *6*, 305–316. [CrossRef] [PubMed]
96. Cohen, A.E.; Soltis, S.M.; González, A.; Aguila, L.; Alonso-Mori, R.; Barnes, C.O.; Baxter, E.L.; Brehmer, W.; Brewster, A.S.; Brunger, A.T.; et al. Goniometer-Based Femtosecond Crystallography with X-Ray Free Electron Lasers. *Proc. Natl. Acad. Sci. USA* **2014**, *111*, 17122–17127. [CrossRef] [PubMed]
97. Barends, T.R.M.; Gorel, A.; Bhattacharyya, S.; Schirò, G.; Bacellar, C.; Cirelli, C.; Colletier, J.P.; Foucar, L.; Grünbein, M.L.; Hartmann, E.; et al. Influence of Pump Laser Fluence on Ultrafast Myoglobin Structural Dynamics. *Nature* **2024**, *626*, 905–911. [CrossRef] [PubMed]
98. Schirò, G.; Woodhouse, J.; Weik, M.; Schlichting, I.; Shoeman, R.L. Simple and Efficient System for Photoconverting Light-Sensitive Proteins in Serial Crystallography Experiments. *J. Appl. Crystallogr.* **2017**, *50*, 932–939. [CrossRef]
99. Fadini, A.; Hutchison, C.D.M.; Morozov, D.; Chang, J.; Maghlaoui, K.; Perrett, S.; Luo, F.; Kho, J.C.X.; Romei, M.G.; Morgan, R.M.L.; et al. Serial Femtosecond Crystallography Reveals That Photoactivation in a Fluorescent Protein Proceeds via the Hula Twist Mechanism. *J. Am. Chem. Soc.* **2023**, *145*, 15796–15808. [CrossRef]
100. Kern, J.; Tran, R.; Alonso-Mori, R.; Koroidov, S.; Echols, N.; Hattne, J.; Ibrahim, M.; Gul, S.; Laksmono, H.; Sierra, R.G.; et al. Taking Snapshots of Photosynthetic Water Oxidation Using Femtosecond X-Ray Diffraction and Spectroscopy. *Nat. Commun.* **2014**, *5*, 4371. [CrossRef] [PubMed]
101. Hattne, J.; Echols, N.; Tran, R.; Kern, J.; Gildea, R.J.; Brewster, A.S.; Alonso-Mori, R.; Glöckner, C.; Hellmich, J.; Laksmono, H.; et al. Accurate Macromolecular Structures Using Minimal Measurements from X-Ray Free-Electron Lasers. *Nat. Methods* **2014**, *11*, 545–548. [CrossRef]
102. Samarkina, O.N.; Popova, A.G.; Gvozdik, E.Y.; Chkalina, A.V.; Zvyagin, I.V.; Rylova, Y.V.; Rudenko, N.V.; Lusta, K.A.; Kelmanson, I.V.; Gorokhovatsky, A.Y.; et al. Universal and Rapid Method for Purification of GFP-like Proteins by the Ethanol Extraction. *Protein Expr. Purif.* **2009**, *65*, 108–113. [CrossRef]
103. Dümmler, A.; Lawrence, A.M.; de Marco, A. Simplified Screening for the Detection of Soluble Fusion Constructs Expressed in *E. coli* using a modular set of vectors. *Microb. Cell Factories* **2005**, *4*, 34. [CrossRef]
104. Van Den Berg, S.; Löfdahl, P.Å.; Härd, T.; Berglund, H. Improved Solubility of TEV Protease by Directed Evolution. *J. Biotechnol.* **2006**, *121*, 291–298. [CrossRef]
105. Schmidt, M.; Pandey, S.; Mancuso, A.; Beam, R. Time-Resolved Serial Femtosecond Crystallography at the European X-Ray Free Electron Laser. *Protocol Exchange* **2019**, *1*. [CrossRef]
106. Kim, Y.; Ganesan, P.; Ihee, H. High-Throughput Instant Quantification of Protein Expression and Purity Based on Photoactive Yellow Protein Turn off/on Label. *Protein Sci.* **2013**, *22*, 1109–1117. [CrossRef]
107. Sikorski, M.; Ramilli, M.; de Wijn, R.; Hinger, V.; Mozzanica, A.; Schmitt, B.; Han, H.; Bean, R.; Bielecki, J.; Bortel, G.; et al. First Operation of the JUNGFRAU Detector in 16-Memory Cell Mode at European XFEL. *Front. Phys.* **2023**, *11*, 1303247. [CrossRef]
108. Kirkwood, H.J.; de Wijn, R.; Mills, G.; Letrun, R.; Kloos, M.; Vakili, M.; Karnevskiy, M.; Ahmed, K.; Bean, R.J.; Bielecki, J.; et al. A Multi-Million Image Serial Femtosecond Crystallography Dataset Collected at the European XFEL. *Sci. Data* **2022**, *9*, 161. [CrossRef] [PubMed]
109. Kruglik, S.G.; Yoo, B.K.; Franzen, S.; Vos, M.H.; Martin, J.L.; Negrerie, M. Picosecond Primary Structural Transition of the Heme Is Retarded after Nitric Oxide Binding to Heme Proteins. *Proc. Natl. Acad. Sci. USA* **2010**, *107*, 13678–13683. [CrossRef]

110. Mizutani, Y.; Kitagawa, T. Ultrafast Dynamics of Myoglobin Probed by Time-Resolved Resonance Raman Spectroscopy. *Chem. Rec.* **2001**, *1*, 258–275. [CrossRef]
111. Levantino, M.; Lemke, H.T.; Schirò, G.; Glownia, M.; Cupane, A.; Cammarata, M. Observing Heme Doming in Myoglobin with Femtosecond X-Ray Absorption Spectroscopy. *Struct. Dyn.* **2015**, *2*, 041713. [CrossRef]
112. Larsen, D.S.; Vengris, M.; Van Stokkum, I.H.M.; Van Der Horst, M.A.; De Weerd, F.L.; Hellingwerf, K.J.; Van Grondelle, R. Photoisomerization and Photoionization of the Photoactive Yellow Protein Chromophore in Solution. *Biophys. J.* **2004**, *86*, 2538–2550. [CrossRef]
113. Nakamura, R.; Hamada, N. Vibrational Energy Flow in Photoactive Yellow Protein Revealed by Infrared Pump-Visible Probe Spectroscopy. *J. Phys. Chem. B* **2015**, *119*, 5957–5961. [CrossRef] [PubMed]
114. Baca, M.; Borgstahl, G.E.O.; Boissinot, M.; Burke, P.M.; Williams, D.W.R.; Slater, K.A.; Getzoff, E.D. Complete Chemical Structure of Photoactive Yellow Protein: Novel Thioester-Linked 4-Hydroxycinnamyl Chromophore and Photocycle Chemistry. *Biochemistry* **1994**, *33*, 14369–14377. [CrossRef]
115. Genick, U.K.; Devanathan, S.; Meyer, T.E.; Canestrelli, I.L.; Williams, E.; Cusanovich, M.A.; Tollin, G.; Getzoff, E.D. Active Site Mutants Implicate Key Residues for Control of Color and Light Cycle Kinetics of Photoactive Yellow Protein. *Biochemistry* **1997**, *36*, 8–14. [CrossRef] [PubMed]
116. Thomson, B.; Both, J.; Wu, Y.; Parrish, R.M.; Martínez, T.J.; Boxer, S.G. Perturbation of Short Hydrogen Bonds in Photoactive Yellow Protein via Noncanonical Amino Acid Incorporation. *J. Phys. Chem. B* **2019**, *123*, 4844–4849. [CrossRef]
117. Schulz, J.; Bielecki, J.; Doak, R.B.; Dörner, K.; Graceffa, R.; Shoeman, R.L.; Sikorski, M.; Thute, P.; Westphal, D.; Mancuso, A.P. A Versatile Liquid-Jet Setup for the European XFEL. *J. Synchrotron Radiat.* **2019**, *26*, 339–345. [CrossRef]
118. Han, H.; Round, E.; Schubert, R.; Gül, Y.; Makroczyová, J.; Meza, D.; Heuser, P.; Aepfelbacher, M.; Barák, I.; Betzel, C.; et al. The XBI BioLab for Life Science Experiments at the European XFEL. *J. Appl. Crystallogr.* **2021**, *54*, 7–21. [CrossRef]

biomolecules

MDPI

Review

AFM for Studying the Functional Activity of Enzymes

Irina A. Ivanova, Anastasia A. Valueva, Maria O. Ershova and Tatiana O. Pleshakova *

Institute of Biomedical Chemistry, Pogodinskaya Str., 10, 119121 Moscow, Russia; i.a.ivanova@bk.ru (I.A.I.); varuevavarueva@gmail.com (A.A.V.); motya00121997@mail.ru (M.O.E.)
* Correspondence: topleshakova@yandex.ru

Abstract: The conventional approach to investigating enzyme systems involves the simultaneous investigation of a large number of molecules and observing ensemble-averaged properties. However, modern science allows us to study the properties of single molecules and to obtain data on biochemical systems at a fundamentally new level, significantly expanding our understanding of the mechanisms of biochemical processes. Imaging of single biomolecules with high spatial and temporal resolution is among such modern research tools. To effectively image the individual steps or intermediates of biochemical reactions in single-molecule experiments, we need to develop a methodology for data acquisition and analysis. Its development will make it possible to solve the problem of separating the static and dynamic disorder present in the parameters identified by traditional proteomic methods. Such a methodology may be based on AFM imaging, the high-resolution microscopic visualization of enzymes. This review focuses on this direction of research, including the relevant methodological and practical solutions related to the potential of developing a single-molecule approach to the study of enzyme systems using AFM-based techniques. We focus on the results of enzyme reaction studies, as there are still few such studies, as opposed to the AFM studies of the mechanical properties of individual enzyme molecules.

Keywords: atomic force microscopy; single-molecule enzymology; protein structure; protein function; AFM imaging

1. Introduction

The study of kinetic processes at the single-molecule level is a relatively recent direction in modern biochemistry. According to statistics from the PubMed platform, more than 400 papers on this topic (retrieved using the keyword string "single-molecule enzymology") have been published worldwide in the last five years, which is an indication of its relevance. This is a promising trend due to the potential to use the resulting knowledge to diagnose diseases associated with enzyme dysfunction [1,2]. Research into kinetic activity lies at the interface of different domains of science, from classical enzymology to single-molecule biophysics. The key advantage of using biophysical methods to study kinetic processes at the single-molecule level is that they allow one to investigate the heterogeneity of free energy states in molecular populations, which is generally a challenging problem for conventional ensemble averaging approaches [3]. Brownian fluctuations and thermal forces are factors playing an important role in molecular heterogeneity. They are the main source of noise and variability in single-molecule experiments. In many cases, it is difficult to differentiate between molecular heterogeneity and stochastic noise caused by thermal effects. A system can be characterized by both temporal and spatial heterogeneity; single-molecule studies allow these differences to be tracked. In cases when the ensemble methods allow one to obtain mean values only, the techniques enabling single-molecule detection may eventually provide not only the mean value but also the probability distribution on both sides of this mean value. The distribution of values with distinct clusters separated by gaps may indicate different energy or conformational states, while the position of the mean value relative to two distinct clusters may indicate that a certain state is preferred over another. The probability distribution of values may detect deviations from the statistical mean behavior and, with sufficient biological and physical insight, become a basis for investigating the potential mechanism of the observed behavior that lies outside the scope of what can be deduced from the simple average value obtained using the integral method [4].

Single-molecule studies of enzyme systems allow the observation of transient states and intermediate products for which information can be lost during ensemble measurements. It is possible to determine the dependence between the mechanistic movements of a globule and the catalytic function within a single molecule. It is clear that in enzymatic reactions for which the activity is measured using the conventional ensemble method (e.g., as the number of enzymatic cycles per unit of time), the functions of the molecules that make up the system as a whole are not synchronized. In other words, at each specific moment, each enzyme molecule can undergo a different stage of the reaction. In the meantime, chemical reactions carried out in a single-molecule system give an idea of the fluctuations in properties that are disguised when the properties of a molecular ensemble are measured [5]. Thus, there are models that allow one to determine the distribution of a property over time in simple extreme cases rather than an averaged value, as well as study their sensitivity to the initial conditions [6]. The biochemical sense of a catalytic reaction at the single-molecule level has been demonstrated by both computational and experimental studies, which have shown that the Michaelis–Menten equation is still valid even for an enzyme but has a different microscopic interpretation [7].

Advances in experimental and computational methods have spurred the emergence of integrated tools that can be used in research related to stubborn biological problems. Experimental progress has been achieved due to the enhancement of the sensitivity and operating speed of sensors, the stability and efficiency of light sources, probes, and microfluidic devices, as well as important factors such as improvements in the mathematical methods underlying the operation of the equipment [4]. Biophysical methods have already been used in single-molecule studies from the perspective of common soft condensed matter [8], as well as complex nanosized biomolecular machines [9] and features of the behavior of individual molecules in living cells [10,11].

The key characteristics of enzyme-catalyzed reactions have been extensively studied at the single-molecule level using a variety of physicochemical methods. Protein structures and functions have been determined, but most of them have been reported for the static molecular conformation. Thus, a lot of information on protein structure has been obtained using techniques such as small-angle X-ray scattering [12] and nuclear magnetic resonance (NMR) [12–14]. The dynamic investigation of the properties of analyzed molecules during reactive events, both in simple single-step reactions and in complex multistep processes, is a relevant problem. A large body of data has been obtained using techniques such as total internal reflection fluorescence microscopy (TIRFM), confocal microscopy, single-molecule fluorescence resonance energy transfer (smFRET), fluorescence correlation spectroscopy (FCS), optical tweezers (OTs), and magnetic tweezers (MTs) [15].

Combinations of methods such as correlative atomic force microscopy and fluorescence microscopy are often used for measurements in environments that are closest to physiological ones. Single-molecule fluorescence analysis has provided valuable data on the correlation between changes in the rate of a catalytic reaction and the conformational fluctuations of the enzyme (e.g., [7]). Fluorescence microscopy techniques that are used for the study of enzymatic reactions lie beyond the scope of this review. For fluorescence-based methods, it is required to insert specific labels in which dyes are covalently linked after site-directed mutagenesis, which affects the native structure of protein molecules. Furthermore, the processes that occur during label staining limit the time range of any particular experiment to three orders of magnitude or less, although single-photon detection has a nanosecond temporal resolution, and fluorescence experiments can technically last for hours [16,17].

The interactions between objects are force-driven, so they are key parameters in biological mechanisms ranging from physiological to cellular and molecular processes such as cotranslational folding, sensory reception, adhesion, and cohesion [18,19]. Processes such as protein folding, translocation, substance transport, and biomolecular interactions involving rearrangements of molecular conformations cannot be observed using the conventional biophysical tools such as nuclear magnetic resonance (NMR) spectroscopy, transmission electron microscopy (TEM), circular dichroism (CD), or any type of fluorescence spectroscopy, since the aforementioned technologies are not intended to apply force to biomolecules, including for the study of elastic properties [20].

Combined molecular dynamics simulations have recently been widely used to study the significance of the specific conformational rearrangements of enzymes during their functioning (for example, [21]). Despite the improvement in modeling methods, important limitations remain. The main limitation is that modeling cannot reach the time scales of most enzymatic reactions. Therefore, it is necessary to develop new experimental methods capable of investigating the dynamics of enzymes during catalysis, providing new information that complements the results obtained using existing experimental approaches and molecular modeling methods.

Contrariwise, the atomic force microscopy (AFM) allows the analysis of molecules by imitating conditions close to the cellular environment, as well as the study of objects in high nanometer-scale resolution [22]. The underlying principle of AFM enables the high-precision control of applied forces and the monitoring of objects at the molecular level. As for enzymes, they are a convenient system for observing the functional properties of single-protein molecules, since most of them undergo conformational changes during a catalytic reaction. Therefore, AFM enables monitoring the functioning of single-enzyme molecules in a liquid medium (i.e., under conditions maximally close to the native ones).

2. The AFM Principle

The underlying principle of AFM measurements is based on the interaction between the scanning element (a cantilever or probe) and the sample surface on which the analyzed bio-object is adsorbed. The necessary condition for using AFM is the adsorption of the analyzed enzyme onto the atomically smooth surface, which implies that the protein properties on the surface differ from those of a molecule in the solution. However, the immobilized enzyme system is used in many biotechnological, biosensor, and medical diagnostic systems [23]. From the perspective of conformational characteristics and retaining the functional properties of biomolecules, it might seem that the need to use the surface is a drawback of AFM. However, numerous studies reporting the results of using such systems to solve bioassay-related problems [24] confirm the stability of immobilized biomolecules and the preservation of their structural and functional properties [25,26].

As mentioned above, in AFM, the surface is scanned using an AFM probe; in most commercial AFM probes, the tip has the shape of a pointed pyramid. The size of the sensitive zone of the probe (the tip curvature radius) is several nanometers, which is comparable to the size of most globular proteins, including enzymes. According to TEM, NMR, and X-ray data, the protein size ranges from 5 to 10 nm [27], while the minimum radius of curvature of the probe tip is 1 nm, providing high-resolution AFM imaging of the surface with the adsorbed protein.

The tapping mode is traditionally used for biological objects. This is a measurement mode based on the detection of changes in the characteristics of probe vibrations depending on the surface topography, usually used for the visualization of fragile objects such as proteins [28]. This measurement mode is preferred because it minimizes the impact of the probe on analyzed objects, preserving their structural and functional properties. It is also a mode that provides new structural and mechanical information about the enzymes being studied. Thus, when using the harmonic oscillator model, the tapping mode based on nonlinear properties can enhance the processes that are barely detectable in the static (contact) mode [28].

The advances in AFM are the result of the optimization of sample preparation methods [29–31] and imaging [32], as well as the continued mastery of the methodology and hardware [33]. At its early stages, AFM analysis could be performed only in the air. More recently, a technique suitable for single-biomolecule analysis in physiologically relevant solutions has been developed, making AFM a sought-after technique in the research into enzyme systems.

The most recent key innovations include the optical positioning system and the bio-AFM liquid cell, AFM based on the quantitative detection of parameters of the force–distance curve (FD-AFM), and high-speed AFM (HS-AFM). Most of these modes are mutually complementary and are used in combinations [34,35]. The AFM methods can traditionally be divided into two large groups: AFM imaging and AFM-based force spectroscopy (AFM-FS). Table 1 summarizes the results of the study of enzyme systems using AFM. Our focus is on the results of studies for enzyme reactions, since there are not many such studies yet, in contrast to the AFM studies of the mechanistic properties of individual enzyme molecules.

Table 1. Summary of the results of the studies of enzyme systems using AFM.

AFM Mode	Enzyme System	Parameter	Reference
AFM-FS	Thioredoxin family	Mechanism of reduction in disulfide bonds	Alegre-Cebollada et al. [36]
	Cellulase (CBM 1, CBH I, and Trichoderma reesei)	The bond strength of individual molecules	Arslan et al. [37]
	Cellulase	Comparison of the adhesion forces between cellulase and lignin with those between cellulase and cellulose and the examination of the moiety groups involved in cellulase binding to lignin	Qin et al. [38]
AFM imaging	Lysozyme	Height fluctuations of lysozyme protein molecules	Radmacher et al. [39]
	P450 CYP102A1	Height fluctuations of protein molecules	Ivanov et al. [40]
	Lipase	Layers degradation induced by the lipase enzyme	Balashev et al. [41]
	Lignin	Layers degradation of lignocellulose films during hydrolysis	Lambert et al. [42]
AFM imaging (HS-AFM)	Caseinolytic peptidase B protein homolog (ClpB)	The dynamics of changes in protein globule morphology and their relationship to catalytic activity	Uchihashi et al. [43]
	ATPase histone chaperone Abo1		Cho et al. [44]
	V1-ATPase		Maruyama et al. [45]
	Laminin-111 and laminin-332		Akter et al. [46]
	Cas9 nuclease		Shibata et al. [47]

3. AFM-Based Force Spectroscopy

AFM-FS techniques applied for mapping the interaction force under different loading conditions are often referred to as dynamic force spectroscopy [48], which is a separate research area for studying the elastic, adhesion, and denaturation properties of protein molecules.

Most studies focus on single-enzyme molecule systems using AFM-FS that deal with protein folding/unfolding. The principle of plotting force curves is based on monitoring probe deflection and piezoelectric element displacement during the cycle of the AFM probe approach and release with respect to the sample surface [49]. Graphically (Figure 1), the force curves show the applied force as a function of the distance between the probe tip and the sample. The investigation of the dynamic sub-angstrom-scale rearrangements of atoms involved in catalysis is an experimentally challenging problem [36].

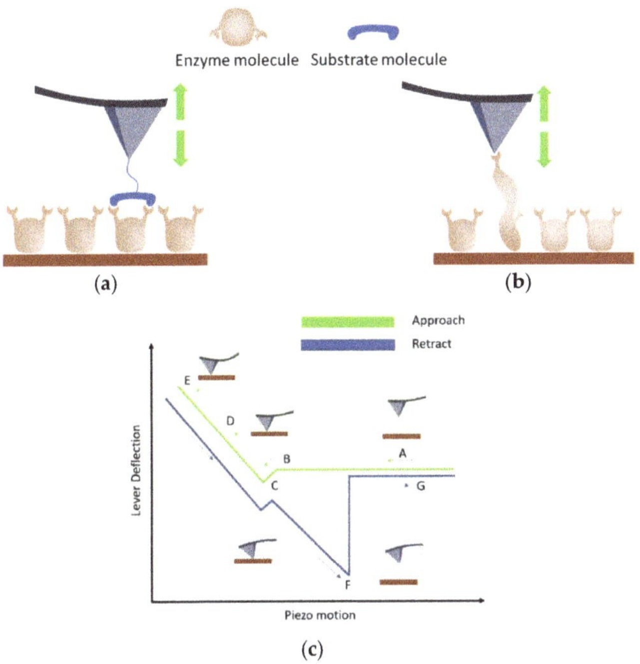

Figure 1. Scheme of data acquisition during measurements in the AFM-FS mode. The probe approaches and is retracted from the surface (green arrows). An enzyme is immobilized on the surface, a substrate molecule is attached to the probe tip (**a**), or the probe is not modified and the "force-clamp" technique is used (**b**). The dependence of the 'lever deflection' on the position of the piezoelectric element is recorded, which ensures the convergence of the probe and the surface. The "lever deflection" signal can be converted into a force value with high accuracy; the recording "piezo-motion" parameters allow us to determine the time of approach and the retraction of the probe from the object. The applied force, probe delay time, approach, and retraction speeds are recorded with high accuracy. A schematic representation of the force–distance curve is shown in (**c**). At large distances (A), the probe and the surface are far from each other, and, therefore, the probe deflection is not measured. As it approaches, the probe begins to feel long-range interactions, mainly of electrostatic and Van der Waals origin (B). The interaction of the surface and the probe tip may be reflected as a jump to the contact (C) in the case of attraction. A further approach leads to the deflection of the probe as it is physically in contact with the surface (D); further compression of the probe on the surface leads to further probe deflection. During the reverse movement of the probe,

hysteresis (F) is observed, since the probe tip interacts with the surface (or enzyme molecule). Depending on the nature of the interaction and the presence of the different force levels of the interactions, the critical adhesion force (F) changes, and characteristic steps appear in the region (C–F). The careful processing of AFM data and a well-prepared experimental design allow obtaining information about intermolecular interactions (in case (**a**)) or about enzyme globule folding/unfolding and conformational rearrangements in the enzyme globule (**b**). Once the probe overcomes the adhesive force, it detaches from the surface (G).

Due to the feasibility of modifying the AFM probe surface and immobilizing molecules on it, AFM-FS can be used to investigate intermolecular interactions in the enzyme system (Figure 1a). In paper [50], the results of the studies of the interaction between lignin and an enzyme are considered. Lignin is a complex polymer that inhibits the enzymatic conversion of cellulose to glucose in lignocellulosic biomass for biofuel production. Cellulase enzymes irreversibly bind to lignin, deactivating the enzyme and reducing the overall activity of the hydrolysis reaction solution [38]. One of the participants in the enzymatic reaction must be attached to the tip of the AFM probe, while the other biomolecules are immobilized on the substrate. The paper [50] summarizes studies using probe-attached receptor molecules to study enzyme–lignin interactions at the single-molecule level.

Conformational rearrangements in the enzyme globule occur during the catalytic cycle due to covalent bond formation/cleavage and atomic rearrangement [51] (Figure 1b). The stiffness of a covalent bond is $\sim$10 nN/Å; the distance in the transition state for a chemical reaction is typically a fraction of an angstrom, so a respective impact of the probe (in the range of ~100 pN to ~1 nN) on the bond is required, which is applicable in the AFM modes [36]. In the modes utilizing a "force clamp" immobilized on the object surface, the force applied to the protein globule can be varied with an accuracy of several hundred piconewtons [52]. The deflection of the probe and the applied force are kept constant by a highly sensitive electronic feedback system [53]. In this approach, forces are applied directly to the disulfide bond in the substrate.

Alegre-Cebollada et al. [36] demonstrated the feasibility of quantifying the effect of the applied force on the enzymatic cleavage of covalent bonds (Figure 2). The results of the application of AFM-FS in the study of the mechanism of disulfide bond reduction by various enzymes of the thioredoxin (Trxs) family are considered. Thioredoxins are reductases that catalyze cysteine–thiol–disulfide exchange reactions. There is evidence that the thioredoxin system is involved in the processes of aging, carcinogenesis, the regulation of proliferation, and apoptosis [54].

The article [36] considers in detail the specific requirements necessary for the application of AFM to an enzyme system. The dependence of the reduction strength of disulfide bonds has been studied for both various chemicals as well as for different Trxs. The work [36] has shown that the enzymatic reaction depends on probe force, which is related to sub-angstrom scale rearrangements in the thioredoxin enzyme and substrate during catalysis [55,56]. To study the process of disulfide bond reduction, two measurement schemes are used in which the applied force is varied. Applying a force of 160–190 pN for 0.3–1.0 s unfolds the domains of the polyprotein. Forces greater than 1 nN are required to break covalent bonds. Therefore, individual unfolding events can be clearly detected as a 10.8 nm step increase in the length of the polyprotein, resulting in well-defined steps in the length-time plot. The shape of the dependences serves as a well-defined fingerprint that definitively distinguishes the polyprotein of interest from any other spurious interactions. Successive unfolding events under the action of force allow the detection of disulfide bonds that have been closed in the protein. The authors showed that the dependence of the reaction rate on the force provides new knowledge into the dynamics of the enzyme and substrate during catalysis. Using the parameter Δx (i.e., the distance to the transition state of the reaction), which can be determined from the obtained force–distance curves, it is possible to assume the presence of the spatial rearrangements of the polypeptide chain regions in the transition state of the reaction. The authors obtained information about the geometry of the transition state by using chemical reagents to cleave the disulfide bond. The authors suggest that the information obtained about the transition state is independent of the lifetime of the enzyme in this state. The advantage is that no matter how short- or long-lived the transition state is, it can always be detected by varying the strength of the impact. The authors emphasize that these rearrangements can be determined on a sub-angstrom scale only using AFM and are unattainable by other modern experimental techniques.

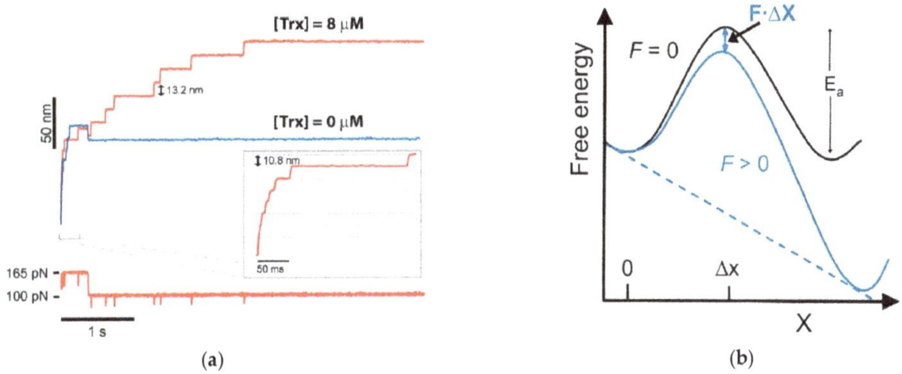

Figure 2. (**a**) At an applied force of 165 pN, a series of 10.8 nm steps is detected, reflecting the rapid unfolding of the modules to the disulfide bond. If a reducing agent such as the Trx enzyme is present in the solution (red curve), a second series of 13.2 nm steps is detected. These correspond to the reduction in disulfide bonds and the subsequent release of residues 32 through 75. In the absence of a reducing agent in solution, no reduction occurs (blue curve) [36]. (**b**) Diagram of the energy landscape for the thiol/disulfide exchange reaction under force, where Δx is the distance to the transition state of the reaction. Figures adapted from [36].

AFM has great potential to study the interaction between lignin and an enzyme under conditions close to physiological ones. Due to the high molecular weight of cellulase, steric hindrances for intermolecular interactions occur when modified on the AFM probe. Most cellulases consist of two domains: the catalytic domain (CD) and the cellulose-binding module (CBM), connected by a highly glycosylated flexible linker. CBM plays an important role in the binding of lignin to cellulase, revealed by molecular dynamics simulations [57]. Since cellulase is not suitable for plate modification, CBM is commonly used to represent cellulase enzymes and to functionalize AFM plates to study the lignin/cellulose-enzyme system at the nanoscale. The authors of this work [50] point out that when studying the force interaction for an enzyme to be immobilized on a probe, the choice of crosslinker is critical. The modification of biomolecules should result in the fact that during the studies, only one enzyme at the tip can contact another molecule modified on the surface [35]. Among the methods, using AFM-FS, which is capable of detecting the bond strength of individual molecules, is often used. Arslan et al. conducted a series of studies [37] on the nanoscale interaction between lignin and CBM (used as a representative enzyme model) (Table 1). It was found that electrostatic and dipole–dipole forces mainly create the interaction between CBM and lignosulfonates. Hydrophobic forces and Lifshitz–van der Waals forces are characteristic of the interaction of kraft lignin and CBM. Organosolv lignin showed the weakest non-productive bond with CBM. Qin et al. [38] studied the interaction between lignin and cellulase based on the adhesion of a substrate-immobilized enzyme to a substrate molecule immobilized on the probe. The immobilization must maintain sufficient mobility and freedom of orientation to facilitate recognition as well as properly bind enzymes to the surface through covalent or noncovalent bonds. The results of this study showed that the measured adhesive forces between lignin and cellulase were, on average, 45% higher than those between hydroxypropyl cellulose and cellulase, demonstrating the specific nature of the binding in the enzyme–substrate complex.

Unfortunately, AFM-FS usually requires longer data acquisition time compared to AFM imaging, thus limiting the wide use of this approach. Moreover, this mode is incapable of detailing a region of the enzyme globule for which the parameter is recorded. As a result, AFM-FS can provide additional information about the mechanism of an enzymatic reaction and changes in the structural properties of a molecule during the reaction at the single-molecule level, but the kinetic parameters of the reaction cannot be determined using this technique.

4. AFM Imaging

The investigation of enzyme systems based on the AFM imaging data (i.e., surface topography measurements) was started quite a long time ago [39]. However, although AFM has been successfully used to visualize protein structures, early studies focusing on the dynamics of changes in protein globules during a reaction were hampered by the low scanning speed of AFM.

The scan rate of standard equipment was, on average, one line per second, so it took up to several minutes to acquire a single image of the desired resolution, whereas biological reactions proceed at the millisecond and shorter time scales. Dynamic processes in protein molecules occur at several time scales: from tens of femtoseconds to hundreds of seconds [58]. Fluctuations or any other conformational motions in proteins are related to time scale (e.g., vibrations occur within hundreds of femtoseconds, while large motions such as domain motions in proteins occur at the millisecond time scale [58]).

Scanning speed is a significant factor affecting the research results and defining the range of problems that can be solved. AFM scanning is accompanied by both sample deformation and image displacement caused by the system drift, especially when scanning in solutions. It is important to take into account data asynchrony, as the data for all the pixels in the image are acquired not simultaneously but with a sequential time delay upon the linewise movement of the probe (for probe scanning) or the specimen stage (for specimen scanning). Accordingly, one can imagine that the slow scanning of a rapidly changing object will result in low spatial resolution [59]. Biological molecules undergo translational diffusion, rotational diffusion, and conformational changes related to their functions, which affect the morphology of the observed objects and, therefore, the apparent spatial resolution of the AFM image. Therefore, an improvement in the temporal resolution of AFM is key to imaging dynamic and mobile biological molecules and clearly resolving their detailed characteristics [60].

In the work [39], the height fluctuations of lysozyme molecules were studied using AFM in the tapping mode in liquid (Figure 3).

The essence of this work [39] is that the height measurements were made by fixing the probe on a lysozyme monolayer without scanning for 32 s. This method made it possible to detect the vibrations of lysozyme molecules in the presence of a substrate in the system at 1 nm, but vibrations were absent in other systems, such as a protein buffer solution, a protein with an inhibitor, and a protein with an inhibitor and a substrate (Figure 4). The authors of the article associate vibrations with possible conformational changes in lysozyme caused by hydrophobic interactions during hydrolysis. It was calculated that a force of 50 pN is required to bend the cantilever by 1 nm, and the energy associated with this bending is approximately 0.1 eV. The enthalpy of hydrolysis is about 0.5 eV per substrate molecule, which is sufficient to bend the cantilever. These measurements indicate that the observed jumps in the measured height in the presence of substrate are probably due to the enzymatic activity of lysozyme.

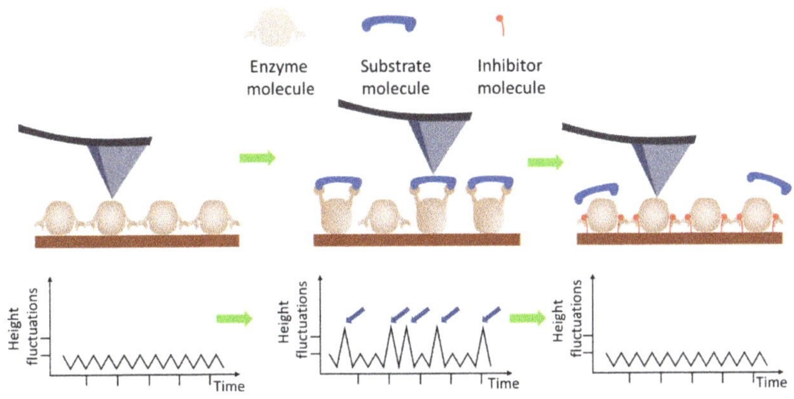

Figure 3. The scheme of measurements in AFM imaging mode to observe protein height oscillations during the catalytic activity of proteins, as proposed in [39]. The probe is installed above a selected surface area containing a layer of enzyme molecules. In the first stage, the height of the objects is recorded under the conditions of an enzyme reaction without the necessary component—the substrate (**left panel**); the initial level of oscillation of the height of the molecule is measured. In the second stage, the substrate is added to the liquid medium, the enzyme reaction is initiated, and the level of oscillation increases (**middle panel**). For control, an inhibitor is added to the system, and the enzyme reaction does not proceed; the level of the height oscillation remains at the background level (**right panel**). These measurements indicate that the observed jumps (**middle panel**, plot image, small blue arrows) in the measured height in the presence of substrate are probably due to the enzymatic activity of lysozyme.

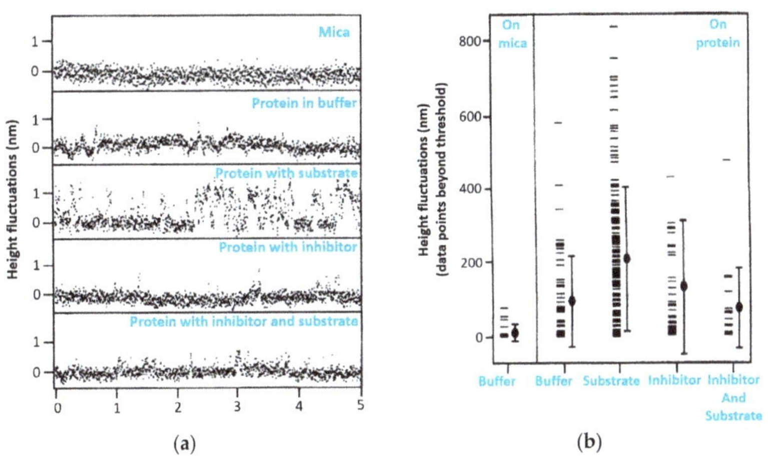

Figure 4. (**a**) Monolayer height variations in lysozyme molecules adsorbed on mica measured by AFM. The data were recorded on mica or on lysozyme in buffer, on lysozyme in buffer containing the substrate 4-methylumbelliferyl-N,N′,N″-triacetyl-chitotriose (~10 μM), in buffer containing the inhibiting substance N,N′-chitobiose (~20 μM), and in buffer containing both substances. Spikelike jumps appear in the height signal. The apparent height of these jumps is on the order of 1 nm. (**b**) Comparison of a dataset of 271 points from six different preparations. The graph shows the points that exceeded the mean value of the data by more than 0.5 nm. The mean and standard deviation are indicated by the filled circle and vertical line. Figure adapted from [39].

In paper [40], cytochrome P450 activity was studied using the measurement principle similar to that reported by Radmacher et al. An approach (Figure 5) has been developed to measure the activity of the single oligomers of the heme-carrying enzyme, cytochrome P450 CYP102A1, based on the AFM imaging data. The amplitude of the height oscillations of single-molecule enzymes involved in the catalytic cycle was shown to be twice as high as the height oscillation amplitude of the same enzymes in the inactive state. It was also demonstrated that the amplitude of the height oscillations of the CYP102A1 protein globule is temperature dependent, and the peak in this curve was observed at 22 °C. The activity of a single CYP102A1 molecule expressed as a unit amplitude of the height oscillations of the protein globule per unit of time was 5 ± 2 Å/s. This process was recorded based on changes in heights (Oz axis, labeled height in schematic Figure 5c) over the observation time for each molecule (Oy axis, labeled time in schematic Figure 5c).

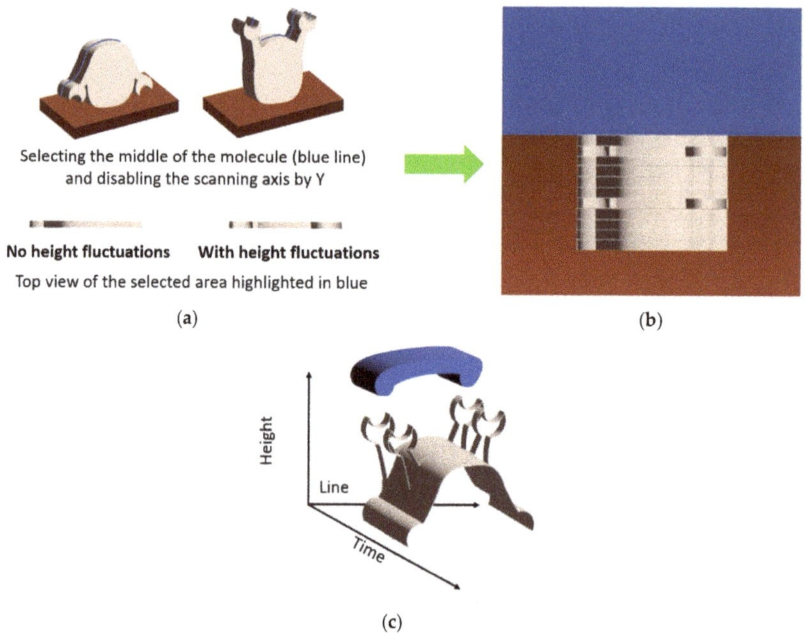

Figure 5. AFM data for the registration of molecular height oscillations during functional activity; an approach was applied in [40]. The height oscillations of individual molecules can be recorded at each stage of the catalytic cycle of the enzyme. (**a**) The first step is the registration of surface topography. The area of interest for the visualized object is selected; (**b**) the second step is topography registration along Ox, and the slow Oy axis was turned off; (**c**) 3D visualization of the second-step results. The Oy axis corresponds to a time of observation for each molecule. The trajectory of the fluctuations of the maximum height of this molecule as a function of time, $h_{max}(t)$, may be determined in the image of the sections of a single molecule.

The kinetics of the enzymatic reaction were also studied by Balashev et al. using the AFM measurements of the topography [41]. This study was based on the evaluation of the occurrence and the growth rate of the defects of the dipalmitoyl phosphatidylcholine (DPPC) degradation layers induced by the lipase enzyme. It was demonstrated that an analysis of sequential AFM images allows one to assess the degradation rate of the adsorbed bilayer, which was attributed to the formation of a product of enzymatic hydrolysis. In this study, the authors relied on an assessment of the changes in the heights of the substrate layers over time, which is not quite suitable for studying the enzyme systems conventionally using low-molecular-weight substances, which are difficult to detect by AFM as substrates.

Also, in our review, it is worth mentioning that one of the articles cited in [50] is devoted to the structural changes in the lignocellulosic substrate in situ during the enzymatic reaction. Lambert et al. proposed using in situ AFM to visually determine the enzymatic hydrolysis of lignocellulosic films with different lignin contents [42]. This paper proposes a strategy to gain insight into the resistance of lignin to enzymatic hydrolysis. Cellulose nanofibril (CNF) lignocellulose films with increased lignin content (up to 40%) were prepared. In situ measurements were performed in real time using atomic force microscopy (AFM) during hydrolysis, and the results were compared with biochemical analyses. Based on the results of the work, the authors emphasized the importance of lignin content and the mutual orientation of CNF and lignin for the efficiency of hydrolysis. An original quantitative analysis of in situ measurements with time-lapse measurements is proposed to visualize the in situ deconstruction of complex lignocellulosic substrates.

A significant breakthrough in the study of biological processes was made as a set of tools has been achieved with the development of a number of tools and the design of high-speed AFM-based (HS-AFM) systems. Several improvements aimed at rapid scanning while focusing on the investigation of biological specimens have been implemented [61–63]. It should be noted that when discussing the speed of HS-AFM imaging, several possible definitions of speed have to be distinguished, such as (a) the image acquisition time (frames/s) and (b) the scanning speed of a probe (m/s). Since we further discuss the use of HS-AFM to study the dynamics of changes in protein globule morphology and their relationship to catalytic activity, the lag time between image acquisitions is more significant than the scanning speed of a probe [64].

The use of small probes with resonant frequencies lying in the megahertz range was one of the key technical solutions [65]. The design of rigid and compact piezoelectric scanners in combination with the development of control techniques has significantly improved the AFM imaging technology [66]. The HS-AFM equipment currently allows scanning at a frame rate of ~33 frames/s and temporal and spatial resolutions comparable to the dynamic analysis of biospecimens [67,68]. The details of the setup and the improvement of its components have recently been thoroughly reviewed in detail in [69,70].

HS-AFM imaging is an approach that has enabled the real-time visualization of biological macromolecules during their functioning. The temporal resolution is typically less than 100 ms; the lateral and vertical spatial resolutions are 2–3 nm and $\sim$0.1 nm, respectively [60]. However, the level of detail obtained in HS-AFM experiments is critically dependent on the spatial and temporal resolutions of the system. HS-AFM has been used to directly observe reactions, such as the formation of the enzyme–substrate complex and protein folding by chaperones, to study caseinolytic peptidase B protein homolog (ClpB) [43], ATPase histone chaperone Abo1 [44], and V1-ATPase [45]. HS-AFM made it possible to detect conformational changes in a molecule upon visualization in solutions, but it is worth mentioning that all the proteins were well immobilized on the mica surface. The choice of an immobilization method is of crucial importance for imaging molecules while preserving their actual structural and functional features and not preventing their conformational changes.

An example of employing HS-AFM is the study of the structural dynamics of laminin-111 and laminin-332 under physiological conditions [46]. The results demonstrate that the coiled-coil domain of laminin-332 is highly dynamically bent around a defined central molecular hinge, whereas the coiled-coil domains of laminin-111 retain their relatively stable S-shaped configuration. Furthermore, structural fluctuations in the cluster of C-terminal LG domains of laminin-111 and laminin-332 between the compact and open conformations were detected, which may play a role in the regulation of the binding of adhesion receptors. Thus, it was demonstrated that HS-AFM can reveal isoform-specific conformational changes occurring in different laminin domains, thus providing new insights into the dynamic structure and specific assembly of laminin affecting the functional properties of the chain.

Another example of using HS-AFM is the visualization of target DNA cleavage by the CRISPR/Cas9 complex during the life cycle [47]. The AFM data have provided data about the functions of the CRISPR/Cas9 complex, including complex assembly, target DNA search, chain cleavage, and product release, which have improved the understanding of the mechanism of action of the gene-editing tool. Of particular interest is that this study has demonstrated the oscillation of the Cas9 nuclease HNH domain, which is responsible for site cleavage, upon binding to the target DNA, and showed differentiation between active and stable closed conformations of a single Cas9 molecule and the Cas9–RNA complex on the mica surface (Figure 6).

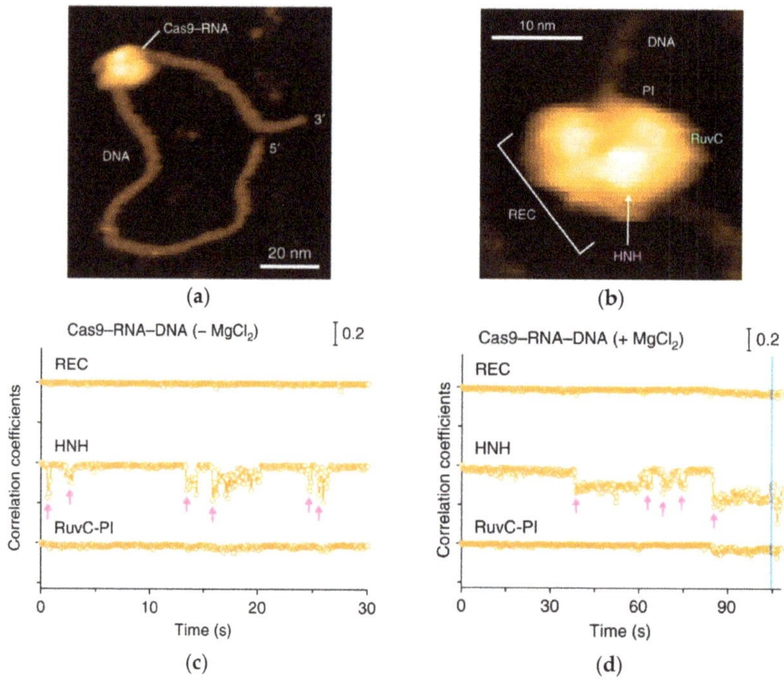

Figure 6. (**a**) Complex assembly of Cas9–RNA–DNA; the scale bar is 20 nm. (**b**) Close-up view of a representative HS-AFM image of Cas9–RNA–DNA. The scale bar is 10 nm. (**c**) Time courses of correlation coefficients for the individual domains between the sequential HS-AFM images of Cas9–RNA–DNA in the absence of $MgCl_2$. The HNH domain fluctuations are indicated by magenta arrows. (**d**) The time courses of the correlation coefficients for the individual domains between the sequential HS-AFM images of Cas9–RNA–DNA in the presence of $MgCl_2$. The HNH domain fluctuations are indicated by magenta arrows. The release of the cleavage product is indicated by a blue line. Figure adapted from [47].

Special attention should be paid to the analysis of the data obtained by high-speed AFM imaging. There exist several methods for analyzing the morphology and motion of a protein globule according to the AFM data (Figure 7). The overall conformational changes for the observed molecules can be determined from their size (e.g., by calculating their circumference), a parameter indicating how circular the contour of the observed molecule is, which is defined as Equation $4\pi S/L2$, where L and S are the contour length and the area enclosed by the contour, respectively [43,71]. When molecule parts change their shape and/or mutual arrangement, conformational changes can be detected by tracking the motion trajectory of a certain section of the moving domain(s)/cluster(s) [68,72]. The histograms of the distances traveled by certain domains allow one to estimate how significant the conformational changes that a molecule undergoes during the observation period are [72]. In all cases, mathematical and statistical analyses based on the observations of a large number of molecules (usually hundreds of molecules) are required to verify the objective conclusion. For example, it has been demonstrated that HS-AFM imaging of single ERdj5 molecules revealed multiple cluster orientations and the highly mobile nature of the C-terminal cluster compared to the N-terminal cluster [72]. The authors analyzed the movement of ERdj5 clusters by tracking trajectories and plotting a histogram of their travel distances. To assess the conformational diversity of ERdj5 molecules, Okumura et al. [73] calculated the circularity for each observed molecule and plotted a histogram. They demonstrated that the average circularity of the wide-type ERdj5 wild type (WT) (0.67) was lower than that for the mutant proteoforms in which the cluster orientation was fixed as form I (0.76) or form II (0.71), and the half-width of the Gaussian fitting curve for WT was the largest among these three variants. This indicated highly dynamic conformational changes for WT and the most circular shape for the mutant proteoform I, as evidenced by the cyclicity revealed by tracking the cluster trajectory [72].

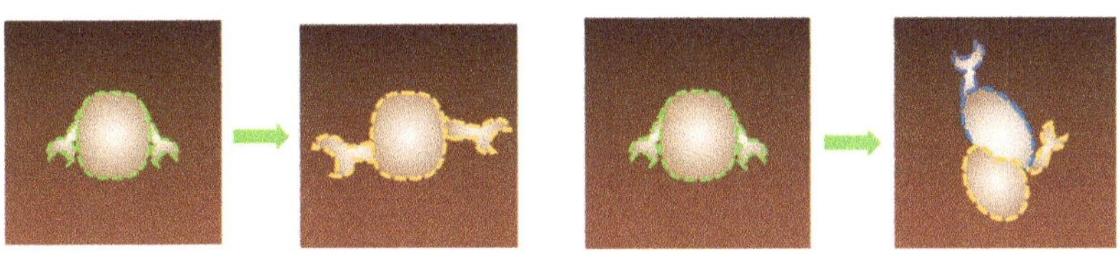

(a) (b)

Figure 7. Scheme of interpretation of data obtained using high-speed AFM imaging (HS-AFM). During the catalytic cycle, a series of images is recorded, the target object—an enzyme molecule or some domain—is isolated, and the morphology of the object is analyzed. There are several methods for analyzing the morphology and motion of a protein globule using AFM data. General conformational changes for the observed molecules can be determined from their size (e.g., by calculating their circumference, a parameter indicating how circular the contour of the observed molecule is) (**a**) or by observing the trajectories of the target structures (**b**).

The analysis of the AFM imaging data acquired by HS-AFM differs significantly from the analysis of the previously used data obtained using the standard equipment. For example, in the study mentioned previously [40], the oscillations of the height of individual CYP102A1 molecules were recorded at each stage of the catalytic cycle of the enzyme during lauric acid hydroxylation using the adapted method described in ref. [39]. The principle of AFM data acquisition is shown in Figure 4. The AFM images of at least ten different molecules of the enzyme were recorded at each stage. First, the surface topography (Figure 5a) was recorded by selecting an object in the image whose height corresponded to that of a protein globule. Then, during AFM visualization, scanning along the slow Oy axis was turned off (Figure 5b). Hence, an image was obtained as "sections" for each visualized molecule (Figure 5c); i.e., a time sweep of one line of a topography image section was recorded. The trajectory of fluctuations of the maximum height of this molecule as a function of time, $h_{max}(t)$, was determined in the image of the sections of a single molecule. For a comparative analysis of the trajectories of different molecules, their linear fitting was performed. The value (the amplitude of height fluctuations for a single molecule over a time interval) was calculated using the result of such an action. Unfortunately, the observation time for each molecule did not exceed 30–40 s. It was not possible to record the signal from a single molecule for a longer period of time, presumably due to substrate drift. Therefore, it was impossible to detect changes in molecule height for a single molecule throughout all the stages of the catalytic cycle, so the data averaged with respect to the results of measuring different molecules at different stages of the catalytic cycle are presented in this study.

In the work [40], measurements were conducted using a Dimension 3100 atomic force microscope (Bruker, Santa Barbara, CA, USA). Today, there is a novel modification of the equipment, the Dimension FastScan atomic force microscope (Bruker, USA), having a unique NanoTrack^{TM} positioning function. The scan area can be positioned and retained for individually selected objects within a single AFM image (Figure 8a). A collection of images is acquired; their analysis for each scan allows one to determine the parameters of a visualized enzyme globule during its functioning. Thus, it is possible to highlight the tip of the molecule (Figure 8b) or an area of interest for the visualized object, which is presumably related to the active site of an enzyme molecule. It is possible to track the maximum height of a protein globule over a long period of time [74] by registering the height of each pixel and determining the ratio between them.

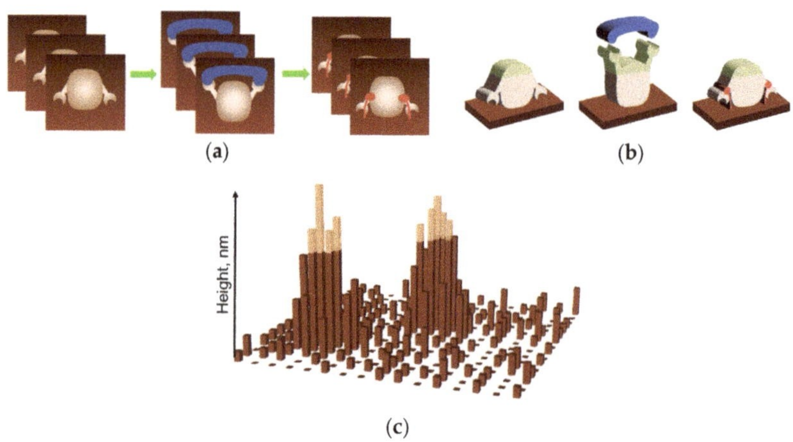

Figure 8. AFM data for recording molecular height variations during functional activity using the NanoTrack^{TM} positioning feature. (**a**) The first step is to register a collection of images for individually selected objects. A collection is created for each catalytic cycle study. (**b**) The second step is to select a region of interest for the visualized object (marked with a green outline). (**c**) The third step is data processing. The height of each pixel and the relationship between them can be determined using a dedicated script [74]. Only the tip of the molecule is considered, and the height of pixels above a certain level (light area in (**c**)) is used for the calculation. It is assumed that the ratio of the height of the pixels in the image of a single molecule can be related to the activity of the molecule during operation. A comparison of data obtained for different catalytic studies allows us to obtain relationships between height and activity at the single-molecule level.

5. Combining AFM with Infrared Spectroscopy

The enhancement of AFM with additional techniques such as infrared spectroscopy (IR) is effectively used to characterize the structural properties of biologically significant materials at the nanoscale. For example, the combination of IR with AFM (IR-AFM) has been successfully used to study amyloid aggregation processes [75], the local broadband spectra of ferritin complexes and insulin aggregates, which can be interpreted in terms of their α-helical and/or β-sheet structure [76].

IR-AFM methods such as scanning Near-Field Scattering Optical Microscopy (s-SNOM) and Peak Force Infrared Microscopy (PFIR) take full advantage of the reduced tip–sample interaction region, achieving a spatial resolution of 10–20 nm without the use of specialized labels. The advantage of these methods is that the spatial resolution is not limited by the diffraction limit (as in classical IR) but by the geometry of the AFM tip and its radius of curvature. In addition to high spatial resolution, the strong near-field enhancement of the tip also provides chemical sensitivity. However, when analyzing complex biological samples containing complexes and groups of biomolecules, the IR spectrum may be a mixture of all the signals obtained [77]. This factor complicates the registration of a signal at the level of single molecules. The isolation of the target spectrum from the general spectrum may require the use of IR labels, which can affect the functional activity of biological objects, in particular, during the catalytic cycle of enzymes. Another disadvantage of IR-AFM methods is that most studies in biological systems have been carried out in the air using dried samples, which limits the space for studying enzymes during the catalytic cycle. It is due to the fact that the implementation of IR-AFM in an aqueous environment is limited to strong signal attenuation in liquid in the mid-IR range [78].

However, the technologies for using IR-AFM in liquids (fundamental, biological, and medicinal) are actively developing and are quite promising. For example, in [79], the authors demonstrated the potential for nanoimaging of s-SNOM in the liquid and conformational identification of catalase nanocrystals, as well as the spatial and spectral analysis of biomimetic peptoid sheets with monolayer sensitivity and chemical specificity at the few-zeptomoles level by additional plasmon field enhancement using metal nanoplates. Also, the successful implementation of the IR-AFM method in liquid is described in [80], where nano-FTIR spectroscopic imaging in liquid was experimentally demonstrated.

In addition, the development of IR-AFM methods can be facilitated by combining mid-IR spectroscopy with high-speed AFM. Since HS-AFM usually uses a shortened cantilever with high resonant frequencies (for example, in MHz), this can potentially contribute to improving the resolution. It is expected that the chemical visualization of the rapid changes in enzyme function in a physiological environment can be successfully implemented and will help to provide additional data for the study of enzyme systems.

It is possible that the development of new technologies in the field of IR-AFM in liquids, will lead to more work being devoted directly to the study of enzymes during the catalytic cycle at the level of individual molecules.

6. Conclusions

The results obtained by various research groups confirm the possibility of using AFM for the visualization, functional, and mechanistic properties of enzymes. Depending on the task, different measurement modes can be used. The need to immobilize the target objects on the surface limits but does not fully restrain the scope of this technique. The results of many studies show that the functional properties of molecules on the surface are preserved. Moreover, the immobilized enzyme system in some cases can become an object of study, since it is used in biosensors or biotechnological workflows. To date, there are not many examples of studying enzyme systems, i.e., the parameters of kinetic processes and not the properties of individual enzyme molecules. Particular attention is paid to the use of HS-AFM, but this equipment, unfortunately, is quite unique and complex. However, this review has shown that standard equipment can be successfully used to study enzyme systems. Understanding the mechanism at a new level of individual molecules is also very important for the development of the industry of producing synthesized enzymes, which are intended to reduce the cost of biotechnological processes. But, as a rule, synthesized enzymes have lower activity than natural analogs. Perhaps the answer to this question can be obtained using AFM as well.

Author Contributions: Conceptualization, T.O.P. and I.A.I.; writing—original draft preparation, I.A.I., A.A.V., and M.O.E.; writing—review and editing, T.O.P. All authors have read and agreed to the published version of the manuscript.

Funding: This work was financed by the Ministry of Science and Higher Education of the Russian Federation within the framework of Agreement No. 075-15-2024-643.

Institutional Review Board Statement: Not applicable.

Informed Consent Statement: Not applicable.

Data Availability Statement: Not applicable.

Conflicts of Interest: The authors declare no conflicts of interest.

References

1. Saghatelian, A.; Cravatt, B. Assignment of protein function in the postgenomic era. *Nat. Chem. Biol.* **2005**, *1*, 130–142. [CrossRef] [PubMed]
2. Qaradakhi, T.; Gadanec, L.K.; McSweeney, K.R.; Tacey, A.; Apostolopoulos, V.; Levinger, I.; Zulli, A. The potential actions of angiotensin-converting enzyme II (ACE2) activator diminazene aceturate (DIZE) in various diseases. *Clin. Exp. Pharmacol. Physiol.* **2020**, *47*, 751–758. [CrossRef]
3. Leake, M.C. The Physics of Life: One Molecule at a Time. *Philos. Trans. R. Soc. B Biol. Sci.* **2013**, *368*, 20120248. [CrossRef]
4. Miller, H.; Zhou, Z.; Shepherd, J.; Wollman, A.J.M.; Leake, M.C. Single-Molecule Techniques in Biophysics: A Review of the Progress in Methods and Applications. *Rep. Prog. Phys.* **2018**, *81*, 024601. [CrossRef]
5. Chen, P.; Zhou, X.; Andoy, N.M.; Han, K.S.; Choudhary, E.; Zou, N.; Shen, H. Spatiotemporal catalytic dynamics within single nanocatalysts revealed by single-molecule microscopy. *Chem. Soc. Rev.* **2014**, *43*, 1107–1117.
6. Margolin, G.; Barkai, E. Single-molecule chemical reactions: Reexamination of the Kramers approach. *Phys. Rev. E* **2005**, *72*, 025101.
7. English, B.P.; Min, W.; van Oijen, A.M.; Lee, K.T.; Luo, G.; Sun, H.; Cherayil, B.J.; Kou, S.C.; Xie, X.S. Ever-Fluctuating Single Enzyme Molecules: Michaelis-Menten Equation Revisited. *Nat. Chem. Biol.* **2006**, *2*, 87–94. [CrossRef]
8. Harriman, O.; Leake, M. Single Molecule Experimentation in Biological Physics: Exploring the Living Component of Soft Condensed Matter One Molecule at a Time. *J. Phys. Condens. Matter Inst. Phys. J.* **2011**, *23*, 503101. [CrossRef]
9. Lenn, T.; Leake, M.C. Experimental Approaches for Addressing Fundamental Biological Questions in Living, Functioning Cells with Single Molecule Precision. *Open Biol.* **2012**, *2*, 120090. [CrossRef]
10. Leake, M.C. Analytical Tools for Single-Molecule Fluorescence Imaging in Cellulo. *Phys. Chem. Chem. Phys.* **2014**, *16*, 12635–12647. [CrossRef]
11. Leake, M. Shining the Spotlight on Functional Molecular Complexes. *Commun. Integr. Biol.* **2010**, *3*, 415–418. [CrossRef] [PubMed]
12. Gomes, G.N.W.; Krzeminski, M.; Namini, A.; Martin, E.W.; Mittag, T.; Head-Gordon, T.; Gradinaru, C.C. Conformational ensembles of an intrinsically disordered protein consistent with NMR, SAXS, and single-molecule FRET. *J. Am. Chem. Soc.* **2020**, *142*, 15697–15710. [CrossRef] [PubMed]
13. Akutsu, H. Strategies for Elucidation of the Structure and Function of the Large Membrane Protein Complex, FoF1-ATP Synthase, by Nuclear Magnetic Resonance. *Biophys. Chem.* **2023**, *296*, 106988. [CrossRef]
14. Wagner, G.; Wüthrich, K. Dynamic model of globular protein conformations based on NMR studies in solution. *Nature* **1978**, *275*, 247–248. [CrossRef]
15. Wang, H.; Zhu, C.; Li, D. Visualizing Enzyme Catalytic Process Using Single-Molecule Techniques. *TrAC Trends Anal. Chem.* **2023**, *163*, 117083. [CrossRef]
16. Lu, M.; Ma, X.; Castillo-Menendez, L.R.; Gorman, J.; Alsahafi, N.; Ermel, U.; Terry, D.S.; Chambers, M.; Peng, D.; Zhang, B.; et al. Associating HIV-1 Envelope Glycoprotein Structures with States on the Virus Observed by smFRET. *Nature* **2019**, *568*, 415–419. [CrossRef] [PubMed]
17. Schmid, S.; Hugel, T. Controlling Protein Function by Fine-Tuning Conformational Flexibility. *eLife* **2020**, *9*, e57180. [CrossRef] [PubMed]
18. Maître, J.-L.; Heisenberg, C.-P. The Role of Adhesion Energy in Controlling Cell–Cell Contacts. *Curr. Opin. Cell Biol.* **2011**, *23*, 508–514. [CrossRef]

19. Wruck, F.; Katranidis, A.; Nierhaus, K.H.; Büldt, G.; Hegner, M. Translation and Folding of Single Proteins in Real Time. *Proc. Natl. Acad. Sci. USA* **2017**, *114*, E4399–E4407. [CrossRef]
20. Banerjee, S.; Chakraborty, S.; Sreepada, A.; Banerji, D.; Goyal, S.; Khurana, Y.; Haldar, S. Cutting-Edge Single-Molecule Technologies Unveil New Mechanics in Cellular Biochemistry. *Annu. Rev. Biophys.* **2021**, *50*, 419–445. [CrossRef]
21. Bhatt, P.; Joshi, T.; Bhatt, K.; Zhang, W.; Huang, Y.; Chen, S. Binding Interaction of Glyphosate with Glyphosate Oxidoreductase and C–P Lyase: Molecular Docking and Molecular Dynamics Simulation Studies. *J. Hazard. Mater.* **2021**, *409*, 124927. [CrossRef] [PubMed]
22. Wu, W.-Q.; Zhu, X.; Song, C.-P. Single-Molecule Technique: A Revolutionary Approach to Exploring Fundamental Questions in Plant Science. *New Phytol.* **2019**, *223*, 508–510.
23. Sheldon, R.A.; Woodley, J.M. Role of biocatalysis in sustainable chemistry. *Chem. Rev.* **2018**, *118*, 801–838. [CrossRef]
24. López_Marzo, A.M. Techniques for Characterizing Biofunctionalized Surfaces for Bioanalysis Purposes. *Biosens. Bioelectron.* **2024**, *263*, 116599. [CrossRef] [PubMed]
25. Josephs, E.A.; Ye, T. Nanoscale spatial distribution of thiolated DNA on model nucleic acid sensor surfaces. *ACS Nano* **2013**, *7*, 3653–3660. [CrossRef] [PubMed]
26. Josephs, E.A.; Ye, T. A single-molecule view of conformational switching of DNA tethered to a gold electrode. *J. Am. Chem. Soc.* **2012**, *134*, 10021–10030.
27. Branden, C.I.; Tooze, J. *Introduction to Protein Structure*, 2nd ed.; Garland Science: New York, NY, USA, 1998; ISBN 978-0-429-06209-4.
28. Ukraintsev, A.A.; Kutuzov, M.M.; Lavrik, O.I. Studying Structure and Functions of Nucleosomes with Atomic Force Microscopy. *Biochem. Mosc.* **2024**, *89*, 674–687. [CrossRef]
29. Mou, J.; Yang, J.; Shao, Z. Atomic Force Microscopy of Cholera Toxin B-Oligomers Bound to Bilayers of Biologically Relevant Lipids. *J. Mol. Biol.* **1995**, *248*, 507–512. [CrossRef]
30. Müller, D.J.; Amrein, M.; Engel, A. Adsorption of Biological Molecules to a Solid Support for Scanning Probe Microscopy. *J. Struct. Biol.* **1997**, *119*, 172–188. [CrossRef]
31. Czajkowsky, D.M.; Sheng, S.; Shao, Z. Staphylococcal α-Hemolysin Can Form Hexamers in Phospholipid Bilayers. *J. Mol. Biol.* **1998**, *276*, 325–330. [CrossRef]
32. Müller, D.J.; Fotiadis, D.; Scheuring, S.; Müller, S.A.; Engel, A. Electrostatically Balanced Subnanometer Imaging of Biological Specimens by Atomic Force Microscope. *Biophys. J.* **1999**, *76*, 1101–1111. [CrossRef] [PubMed]
33. Hansma, P.K.; Cleveland, J.P.; Radmacher, M.; Walters, D.A.; Hillner, P.E.; Bezanilla, M.; Fritz, M.; Vie, D.; Hansma, H.G.; Prater, C.B.; et al. Tapping Mode Atomic Force Microscopy in Liquids. *Appl. Phys. Lett.* **1994**, *64*, 1738–1740. [CrossRef]
34. Alsteens, D.; Newton, R.; Schubert, R.; Martinez-Martin, D.; Delguste, M.; Roska, B.; Müller, D.J. Nanomechanical Mapping of First Binding Steps of a Virus to Animal Cells. *Nat. Nanotechnol.* **2017**, *12*, 177–183. [CrossRef] [PubMed]
35. Liang, W.; Shi, H.; Yang, X.; Wang, J.; Yang, W.; Zhang, H.; Liu, L. Recent Advances in AFM-Based Biological Characterization and Applications at Multiple Levels. *Soft Matter* **2020**, *16*, 8962–8984. [CrossRef]
36. Alegre-Cebollada, J.; Perez-Jimenez, R.; Kosuri, P.; Fernandez, J.M. Single-Molecule Force Spectroscopy Approach to Enzyme Catalysis. *J. Biol. Chem.* **2010**, *285*, 18961–18966. [CrossRef]
37. Arslan, B.; Colpan, M.; Ju, X.; Zhang, X.; Kostyukova, A.; Abu-Lail, N.I. The Effects of Noncellulosic Compounds on the Nanoscale Interaction Forces Measured between Carbohydrate-Binding Module and Lignocellulosic Biomass. *Biomacromolecules* **2016**, *17*, 1705–1715. [CrossRef]
38. Qin, C.; Clarke, K.; Li, K. Interactive forces between lignin and cellulase as determined by atomic force microscopy. *Biotechnol. Biofuels* **2014**, *7*, 65. [CrossRef]
39. Radmacher, M.; Fritz, M.; Hansma, H.; Hansma, P.K. Direct Observation of Enzyme Activity with the Atomic Force Microscope. *Science* **1994**, *265*, 1577–1579. [CrossRef]
40. Ivanov, Y.D.; Bukharina, N.S.; Pleshakova, T.O.; Frantsuzov, P.A.; Krokhin, N.V.; Ziborov, V.S.; Archakov, A.I. Atomic Force Microscopy Visualization and Measurement of the Activity and Physicochemical Properties of Single Monomeric and Oligomeric Enzymes. *Biophysics* **2011**, *56*, 892–896. [CrossRef]
41. Balashev, K.; Nielsen, L.K.; Callisen, T.; Svendsen, A. In Situ Studies of Single Enzymes and Enzyme Kinetics by Atomic Force Microscopy (AFM). *Probe Microsc.* **2001**, *2*, 177–185.
42. Lambert, E.; Aguié-Béghin, V.; Dessaint, D.; Foulon, L.; Chabbert, B.; Paës, G.; Molinari, M. Real time and quantitative imaging of lignocellulosic films hydrolysis by atomic force microscopy reveals lignin recalcitrance at nanoscale. *Biomacromolecules* **2018**, *20*, 515–527. [PubMed]
43. Uchihashi, T.; Watanabe, Y.; Nakazaki, Y.; Yamasaki, T.; Watanabe, H.; Maruno, T.; Ishii, K.; Uchiyama, S.; Song, C.; Murata, K.; et al. Dynamic Structural States of ClpB Involved in Its Disaggregation Function. *Nat. Commun.* **2018**, *9*, 2147. [CrossRef] [PubMed]

44. Cho, C.; Jang, J.; Kang, Y.; Watanabe, H.; Uchihashi, T.; Kim, S.J.; Kato, K.; Lee, J.Y.; Song, J.-J. Structural Basis of Nucleosome Assembly by the Abo1 AAA+ ATPase Histone Chaperone. *Nat. Commun.* **2019**, *10*, 5764. [CrossRef] [PubMed]
45. Maruyama, S.; Suzuki, K.; Imamura, M.; Sasaki, H.; Matsunami, H.; Mizutani, K.; Saito, Y.; Imai, F.L.; Ishizuka-Katsura, Y. Kimura-Someya, T.; et al. Metastable Asymmetrical Structure of a Shaftless V1 Motor. *Sci. Adv.* **2019**, *5*, eaau8149. [CrossRef]
46. Shibata, M.; Nishimasu, H.; Kodera, N.; Hirano, S.; Ando, T.; Uchihashi, T.; Nureki, O. Real-Space and Real-Time Dynamics of CRISPR-Cas9 Visualized by High-Speed Atomic Force Microscopy. *Nat. Commun.* **2017**, *8*, 1430. [CrossRef]
47. Akter, L.; Flechsig, H.; Marchesi, A.; Franz, C.M. Observing Dynamic Conformational Changes within the Coiled-Coil Domain of Different Laminin Isoforms Using High-Speed Atomic Force Microscopy. *Int. J. Mol. Sci.* **2024**, *25*, 1951. [CrossRef]
48. Best, R.B.; Brockwell, D.J.; Toca-Herrera, J.L.; Blake, A.W.; Smith, D.A.; Radford, S.E.; Clarke, J. Force Mode Atomic Force Microscopy as a Tool for Protein Folding Studies. *Anal. Chim. Acta* **2003**, *479*, 87–105. [CrossRef]
49. Cappella, B.; Dietler, G. Force-Distance Curves by Atomic Force Microscopy. *Surf. Sci. Rep.* **1999**, *34*, 1–104. [CrossRef]
50. Zhao, X.; Meng, X.; Ragauskas, A.J.; Lai, C.; Ling, Z.; Huang, C.; Yong, Q. Unlocking the Secret of Lignin-Enzyme Interactions Recent Advances in Developing State-of-the-Art Analytical Techniques. *Biotechnol. Adv.* **2022**, *54*, 107830. [CrossRef]
51. Beyer, M.K.; Clausen-Schaumann, H. Mechanochemistry: The mechanical activation of covalent bonds. *Chem. Rev.* **2005**, *105* 2921–2948. [CrossRef]
52. Schlierf, M.; Li, H.; Fernandez, J.M. The Unfolding Kinetics of Ubiquitin Captured with Single-Molecule Force-Clamp Techniques *Proc. Natl. Acad. Sci. USA* **2004**, *101*, 7299–7304. [CrossRef] [PubMed]
53. Oberhauser, A.F.; Hansma, P.K.; Carrion-Vazquez, M.; Fernandez, J.M. Stepwise Unfolding of Titin under Force-Clamp Atomic Force Microscopy. *Proc. Natl. Acad. Sci. USA* **2001**, *98*, 468–472. [CrossRef]
54. Yoshida, T.; Nakamura, H.; Masutani, H.; Yodoi, J. The involvement of thioredoxin and thioredoxin binding protein-2 on cellular proliferation and aging process. *Ann. New York Acad. Sci.* **2005**, *1055*, 1–12. [CrossRef]
55. Perez-Jimenez, R.; Wiita, A.P.; Rodriguez-Larrea, D.; Kosuri, P.; Gavira, J.A.; Sanchez-Ruiz, J.M.; Fernandez, J.M. Force-Clamp Spectroscopy Detects Residue Co-Evolution in Enzyme Catalysis. *J. Biol. Chem.* **2008**, *283*, 27121–27129. [CrossRef]
56. Wiita, A.; Perez-Jimenez, R.; Walther, K.; Gräter, F.; Berne, B.J.; Holmgren, A.; Sanchez-Ruiz, J.M.; Fernandez, J.M. Probing the chemistry of thioredoxin catalysis with force. *Nature* **2007**, *450*, 124–127. [CrossRef] [PubMed]
57. Vermaas, J.V.; Petridis, L.; Qi, X.; Schulz, R.; Lindner, B.; Smith, J.C. Mechanism of Lignin Inhibition of Enzymatic Biomass Deconstruction. *Biotechnol. Biofuels* **2015**, *8*, 217. [CrossRef]
58. McCammon, J.A.; Harvey, S.C. *Dynamics of Proteins and Nucleic Acids*; Cambridge University Press: Cambridge, UK, 1988.
59. Hall, D.; Foster, A.S. Practical considerations for feature assignment in high-speed AFM of live cell membranes. *Biophys. Physicobiology* **2022**, *19*, e190016.
60. Umeda, K.; McArthur, S.J.; Kodera, N. Spatiotemporal Resolution in High-Speed Atomic Force Microscopy for Studying Biological Macromolecules in Action. *Microscopy* **2023**, *72*, 151–161. [CrossRef]
61. Krishnamurthy, K.; Rajendran, A.; Nakata, E.; Morii, T. Near Quantitative Ligation Results in Resistance of DNA Origami Against Nuclease and Cell Lysate. *Small Methods* **2024**, *8*, 2300999.
62. Fukuda, S.; Ando, T. Technical Advances in High-Speed Atomic Force Microscopy. *Biophys. Rev.* **2023**, *15*, 2045–2058. [CrossRef]
63. Endo, M.; Sugiyama, H. Single-Molecule Visualization of B–Z Transition in DNA Origami Using High-Speed AFM. In *Z-DNA: Methods and Protocols*; Kim, K.K., Subramani, V.K., Eds.; Springer: New York, NY, USA, 2023; pp. 241–250. ISBN 978-1-07-163084-6
64. Rajendran, A.; Endo, M.; Sugiyama, H. State-of-the-Art High-Speed Atomic Force Microscopy for Investigation of Single-Molecular Dynamics of Proteins. *Chem. Rev.* **2014**, *114*, 1493–1520. [CrossRef] [PubMed]
65. Walters, D.A.; Cleveland, J.P.; Thomson, N.H.; Hansma, P.K.; Wendman, M.A.; Gurley, G.; Elings, V. Short cantilevers for atomic force microscopy. *Rev. Sci. Instrum.* **1996**, *67*, 3583–3590.
66. Schitter, G.; Astrom, K.J.; DeMartini, B.E.; Thurner, P.J.; Turner, K.L.; Hansma, P.K. Design and modeling of a high-speed AFM-scanner. *IEEE Trans. Control. Syst. Technol.* **2007**, *15*, 906–915.
67. Kodera, N.; Yamamoto, D.; Ishikawa, R.; Ando, T. Video Imaging of Walking Myosin V by High-Speed Atomic Force Microscopy *Nature* **2010**, *468*, 72–76. [CrossRef] [PubMed]
68. Uchihashi, T.; Iino, R.; Ando, T.; Noji, H. High-Speed Atomic Force Microscopy Reveals Rotary Catalysis of Rotorless F1-ATPase *Science* **2011**, *333*, 755–758. [CrossRef]
69. Ando, T.; Kodera, N.; Uchihashi, T.; Miyagi, A.; Nakakita, R.; Yamashita, H.; Matada, K. High-Speed Atomic Force Microscopy for Capturing Dynamic Behavior of Protein Molecules at Work. *E J. Surf. Sci. Nanotechnol.* **2005**, *3*, 384–392. [CrossRef]
70. Ando, T.; Uchihashi, T.; Fukuma, T. High-Speed Atomic Force Microscopy for Nano-Visualization of Dynamic Biomolecular Processes. *Prog. Surf. Sci.* **2008**, *83*, 337–437. [CrossRef]
71. Okumura, M.; Noi, K.; Kanemura, S.; Kinoshita, M.; Saio, T.; Inoue, Y.; Hikima, T.; Akiyama, S.; Ogura, T.; Inaba, K. Dynamic Assembly of Protein Disulfide Isomerase in Catalysis of Oxidative Folding. *Nat. Chem. Biol.* **2019**, *15*, 499–509. [CrossRef]
72. Maegawa, K.I.; Watanabe, S.; Noi, K.; Okumura, M.; Amagai, Y.; Inoue, M.; Inaba, K. The highly dynamic nature of ERdj5 is key to efficient elimination of aberrant protein oligomers through ER-associated degradation. *Structure* **2017**, *25*, 846–857.

73. Okumura, M.; Noi, K.; Inaba, K. Visualization of Structural Dynamics of Protein Disulfide Isomerase Enzymes in Catalysis of Oxidative Folding and Reductive Unfolding. *Curr. Opin. Struct. Biol.* **2021**, *66*, 49–57. [CrossRef]
74. Ivanova, I.A.; Ershova, M.O.; Pleshakova, T.O. *Data Processing Algorithm for Determining Biomacromolecule Height Fluctuations in AFM Measurements*; Institute of Biomedical Chemistry: Moscow, Russia, 2024; pp. 23–24.
75. Ruggeri, F.S.; Habchi, J.; Cerreta, A.; Dietler, G. AFM-Based Single Molecule Techniques: Unraveling the Amyloid Pathogenic Species. *Curr. Pharm. Des.* **2016**, *22*, 3950–3970. [PubMed]
76. Amenabar, I.; Poly, S.; Nuansing, W.; Hubrich, E.H.; Govyadinov, A.A.; Huth, F.; Krutokhvostov, R.; Zhang, L.; Knez, M.; Heberle, J.; et al. Structural Analysis and Mapping of Individual Protein Complexes by Infrared Nanospectroscopy. *Nat. Commun.* **2013**, *4*, 2890. [CrossRef] [PubMed]
77. VD dos Santos, A.C.; Hondl, N.; Ramos-Garcia, V.; Kuligowski, J.; Lendl, B.; Ramer, G. AFM-IR for Nanoscale Chemical Characterization in Life Sciences: Recent Developments and Future Directions. *ACS Meas. Sci. Au* **2023**, *3*, 301–314. [CrossRef]
78. Ramer, G.; Ruggeri, F.S.; Levin, A.; Knowles, T.P.J.; Centrone, A. Determination of Polypeptide Conformation with Nanoscale Resolution in Water. *ACS Nano* **2018**, *12*, 6612–6619. [CrossRef]
79. O'Callahan, B.T.; Park, K.-D.; Novikova, I.V.; Jian, T.; Chen, C.-L.; Muller, E.A.; El-Khoury, P.Z.; Raschke, M.B.; Lea, A.S. In Liquid Infrared Scattering Scanning Near-Field Optical Microscopy for Chemical and Biological Nanoimaging. *Nano Lett.* **2020**, *20*, 4497–4504. [CrossRef]
80. Virmani, D.; Bylinkin, A.; Dolado, I.; Janzen, E.; Edgar, J.H.; Hillenbrand, R. Amplitude- and Phase-Resolved Infrared Nanoimaging and Nanospectroscopy of Polaritons in a Liquid Environment. *Nano Lett.* **2021**, *21*, 1360–1367. [CrossRef]

MDPI AG
Grosspeteranlage 5
4052 Basel
Switzerland
Tel.: +41 61 683 77 34

Biomolecules Editorial Office
E-mail: biomolecules@mdpi.com
www.mdpi.com/journal/biomolecules